Over 50 Exciting Electronics Experiments

A practical guide for hobbyists and students of electronics engineering

Kommajosyula Krishna Murty

PUSTAK MAHAL®

Publishers
Pustak Mahal®

Administrative office and sale centre

J-3/16 , Daryaganj, New Delhi-110002
☎ 23276539, 23272783, 23272784 • *Fax:* 011-23260518
E-mail: info@pustakmahal.com • *Website:* www.pustakmahal.com

Branches
Bengaluru: ☎ 080-22234025 • *Telefax:* 080-22240209
E-mail: pustakmahalblr@gmail.com
Mumbai: ☎ 022-22010941, 022-22053387
E-mail: unicornbooksmumbai@gmail.com
Patna: ☎ 0612-3294193 • *Telefax:* 0612-2302719
E-mail: rapidexptn@gmail.com

ISBN 978-81-223-1002-3

Edition: 2016

Price : ₹ 195/-

Printed at : Radha Offset, Delhi

DEDICATION

At The Lotus Feet of

Most Gracious Huzur Dr. M.B. Lal Sahab

The Seventh Sant Sadguru Of

Radhasoami Faith (Dayalbagh)

On The Occasion Of

His Birth Centenary Celebrations.

Contents

Preface

Wonder if the trio Shockley, Bardeen and Brattain ever imagined that their invention of a tiny transistor on that fateful day on December 23, 1947 would transform the twentieth century! Have Jack Kilby or Robert Noyce ever thought that their four transistor integrated circuits would one day evolve into chips containing a few million transistors? And Gordon Moore?

As the world is shrinking, boundaries are broken; electronics is smiling from behind in this technological revolution! A single product that managed to alter the course of human history in a short span of time, the transistor!

Hobby in electronics is exciting and its experiments are exhilarating. Electronics 'is a field which grows continuously with ever expanding frontiers.' But.....

Students still wade through a curriculum that laps theory a lot. Practical experience of even a diode evades them. An enthusiastic hobbyist often had to rely on the evasive knowledge of repair mechanics or so called service engineers who learnt their knowledge at the cost of the customers.

In this context, here is a book that abounds with a number of enjoyable electronics projects for upcoming hobbyist and a practical student.

I hear, and I forget.
I see, and I remember.
I do, and I understand.

This old adage generally attributed to Confucius, the Chinese sage, sums up hobby in Electronics.

And to do and understand, here now you are on a guided tour of this interesting hobby.

The engine for this tour should have the power. Hence the travel begins with basic building blocks, the power supplies, leading to simple solder less projects with piezo buzzer.

The next stop is at the irresistible, versatile and very rugged IC, 555 which provides the effortless feel of proper project building.

Then the drive goes through to digital ICs, building alarms for home, automobile and telephone and mains control.

Start with a simple lapel mike in the audio street and go ahead in paces to 20 W(RMS) AMPLIFIER and then to VOICE RECORDING ON CHIP.

Counters and Clock are around but in the end.

In this journey you have constant travel companions with pin outs, truth tables and descriptions of the ICs. Electronic components introduce themselves as they appear with little notes, details and discussion, rather than sit pretty in the pantry car. Unnecessary textual theory which dissuades a normal hobbyist is avoided. As these details are at the project site and not jumbled at a different place, botheration of leafing through the pages back and forth is avoided.

The book begins with simple circuits, goes on to complex circuits. More experienced persons are prompted to skip to the next stop. The trip is practical and useful. No fancy circuits. No obsolete devices. Parts lists are enclosed with the projects. Generally available (even in towns) components are used.

Notes on TIPS AND TRICKS, ART OF SOLDERING AND DESOLDERING, CARE OF ICs, CMOS AND TTL ICs, TROUBLE SHOOTING, etc., is the guard of the train, of course stays at the end.

Details which are available elsewhere are carefully avoided, but at the same time the book is self sustaining and you have practical hints and trouble shooting techniques.

There are two indices, one page wise and the other section wise.

Your driver has over 35 years of practical experience whose expedition in electronics hobby started with good old AC125 germanium transistor. But experience does not preclude mistakes. Welcome!

The journey begins!

—K.KRISHNA MURTY
403, S.B.Enclave, 11-6-2, Rockdale Lay out,
Maharanipeta, Visakhapatnam 530 002
Phone: 91-891 -2547470, Mobile:9440660288
Email:kaykayem@gmail.com, kaykaymurthy@yahoo.com

Power Supplies

Introduction

Most of the present day electronic circuits work on low voltage direct current, whereas the domestic power available in India is high voltage alternating current at 230V. To work on our hobby circuits and general electronic devices, we need to step down this 230V AC to 3, 6, 9, or 12 volts or any other required voltage, and rectify to make it direct current.

A transformer is used to step down the high voltage AC to low voltage AC. Transformer is device with a large number of turns of insulated copper wires wound on iron core, known as Primary side and is connected to AC 230V. A less number of turns are wound over this to give lower voltage, which is known as secondary side. This is known as a step down transformer. Transformers are rated at voltages and current they can deliver. Transformer has an added advantage of isolation, which protects us from shock. After transformation the voltage is still alternating. Hence we have to rectify it to make it direct current for which, we have semiconductor diodes coming in handy.

There are two methods for this rectification process -half wave and full wave. Simple rectification does not yield pure DC; it has still the AC ripple on it. Capacitors filter this out and then we get a reasonably good DC.

Half wave rectifiers have more of this ripple over them and their voltage regulation is poor. Full wave rectifier has better line regulation and has less AC ripple on it because both the half cycles are rectified. But do not underestimate half wave power supply. It is just adequate for many applications and is extensively used. However both these power supplies suffer from the drawback that their voltage drops appreciably as the current drawn from it is increased. These power supplies also fluctuate as the mains power supply fluctuates. When you need to maintain strict stabilized voltage, we have regulated power supplies which keep up constant voltage irrespective of line and load fluctuations.

Here we discuss power supplies from simple half wave rectifier to adjustable regulated ones. Please note that any one of the power supply circuits shown herein are used to power up the later circuits depending on the line, load and circuit requirements. Hence it is good to know a little about them and it is important to build a good power supply. The power supply is the basic building block for our hobby and a beginning is to be made here. On the other hand we have not discussed the latest switch mode power supplies as some parts like high frequency transformers are not easily available for a general hobbyist. In general projects with winding of coils are avoided in this book for the same reason.

Before we go into the actual circuits, let us build a power supply and improve on it. If you are well versed with power supplies and have actually built them, go ahead and skip this section. There are a lot of interesting circuits ahead.

This hobby will be more and more exciting as we dive deeper and deeper into it.

Half Wave Power Supply

Description

Half wave rectification is simplest, as it requires only one transformer, one diode and one capacitor, but we have added a LED to give power supply indication. A simple half wave rectifier for 12V is shown in Schematic 1.

Construction

500mA transformers are generally used for these circuits. Connect the primary wires to the mains chord after carefully insulating the joints. This transformer has three wires on the secondary side. AC voltage across both end wires is 12 and voltage across any one end wire and center wire is 6. So if both end wires are connected as shown in the present schematic the output DC voltage will be 12 and if any one of the end wires and center tap are connected the voltage will be 6.

Take a Veroboard and solder all components on it except mains power supply. Solder IN4003 diode to one end of secondary winding making note of the cathode. Solder 1000uf /25V capacitor. Please note the capacitor is polarized, which means that you should connect it one way only. Negative side of the pin is marked on the can. Add a light emitting diode to know that the power supply is on. Now use a 1K-¼ watt resistor in series with it.

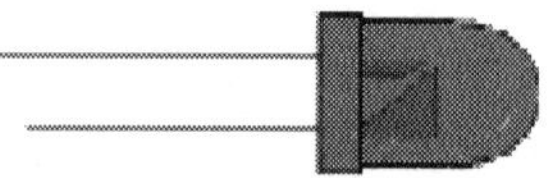

LEDs are also polarized and marked. LEDs will not light up if they are connected in reverse. Take two pen light cells and connect the ends of a Red LED to positive and negative and then turn the leads over and try. (Forget the cathodes and anodes for the time being.) You will know that it works only one way. That also explains how a diode works. Do not connect them directly to 12 V as LEDs cannot work beyond 5 V and their current capability is extremely limited. Although they are very rugged devices, they must always be used with a current limiting resistor.

Parts

Item	*No. reqd*	*Description*	*Designation*
1	1	1000uF	C1
2	1	1N4003	D1
3	1	LED1	D2
4	1	1K	R1
5	1	230V/6-0-6 V Transformer	T1

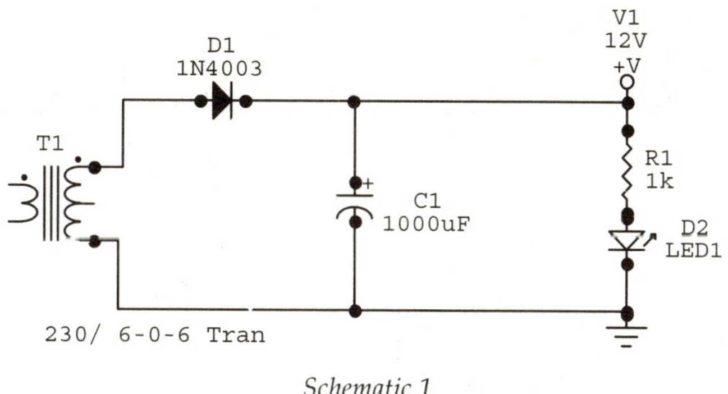

Schematic 1

Full Wave Power Supply

Introduction

Make a full wave rectifier, as it is cheaper and better than a half wave one and has less AC ripple. Full wave rectification can be accomplished by two ways, either by a center-tapped transformer using two diodes, or by a bridge circuit, which uses four diodes.

Method 1

Description

We need center tapped transformer as used in the earlier circuit. .As already described, in this transformer, you have three wires on the secondary side. Simply speaking in this transformer center wire is taken as zero reference; two outside wires give you equal voltages. Say for a 6-0-6 transformer, center wire gives 0 volts while both ends give 6 volts each. The circuit is shown in Schematic 2.

Construction

Connect diodes, capacitor and resistor and LED just as in earlier project but follow the circuit is shown in the Figure below. Now if you wish to use the same 6 – 0 - 6 transformer, you will get 6V DC. If you wish to have 12 V DC by this method, you should get a 12 - 0 -12 transformer.

Solder 1000uf / 25Vcapacitor after carefully noting the polarity. Negative side of the pin is marked on the can. A light emitting diode along with 1K-¼ watt resistor in series with it is added to indicate that the power supply is on.

Parts

ONLY ADDITIONAL PART REQUIRED IS ONE MORE IN4003 DIODE.

Transformer is now rated at 12 - 0 -12V. If you use the same transformer (6-0-6) as in the earlier circuit, you will get 6V DC.

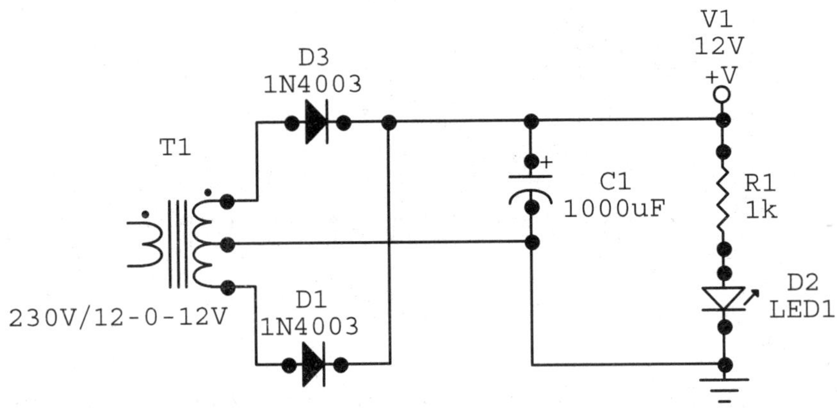

Schematic 2

Method 2

Description

We can still build a full wave rectifier without a center-tapped transformer. We need four diodes for what is known as bridge circuit and a working full wave bridge circuit for 12V is shown in Schematic 3. This is most generally used power supply circuit.

Construction

Follow the same procedure. As you can see the center tap was ignored and two outer ends of the transformer are connected. This gives you 12V DC supply. Transformers without a center tap of the required voltage are available and you can straightaway use it. However if you use center tap and any of the outer taps you will get 6Vr

Solder 1000uf / 25Vcapacitor after carefully noting the polarity. Negative side of the pin is marked on the can. A light emitting diode along with 1K-¼ watt resistor in series with it is added to indicate that the power supply is on.

Parts

ONLY ADDITIONAL PARTS ARE TWO MORE IN4003 DIODES.

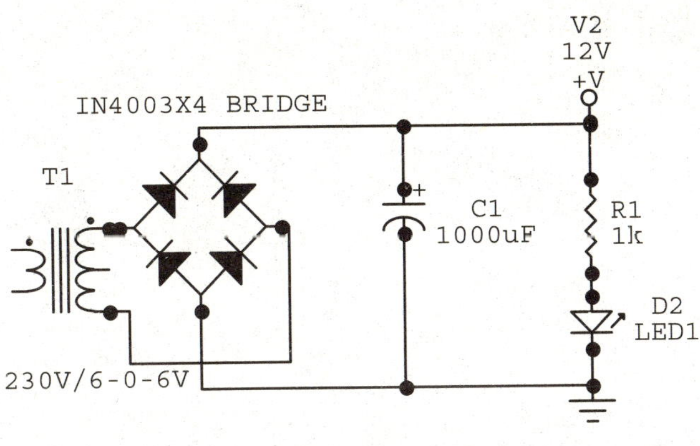

Schematic 3

Let us learn a little about components

Now let us learn a little about some components. You may wonder why these were not discussed in the first place. I felt that this would sound like text book and would dissuade a potential hobbyist in you from turning other leaves of this book. Hence the theory was not discussed in the beginning but was injected into the text where it is required. Wherever text is in italics in this book, it is theory as far as necessary for a hobbyist. Similarly block diagrams, pin out details, functional diagrams are given along side the text with the schematic, rather than bunching them together at a different place. This makes it easy for the project builder to look into all the details at the same place, rather than leafing back and forth. Only objective details that are necessary for a hobbyist are discussed.

Now a word about the diodes

Diode is basically one-way device allowing current to flow only in one direction, which makes it useful for rectification. This was very well observed earlier with LED. IN40- - series diodes are commonly used in these circuits. Without much dwelling deep into the specs and details, let us say that these are small and capable of handling 1 amp current. IN 4001 is capable of handling 50 volts of PIV and IN4007 diode is capable of handling 1200 volts of PIV. Each diode costs less than a rupee.

PIV is peak inverse voltage. Normally these diodes are used with a PIV of at least 6 times the operating voltage. For a 6 volt operation, we may use IN4001 diode but for 230 V operations, we should use IN 4007. These are fairly cheap and cost difference between higher voltage rated diodes to lower ones is not much. Higher rated diodes in this series can be used in low voltage circuits. As far as possible do not cut the leads short. Twist them in to small rings as the lead wires also work as heat sinks. There is a white band marked on the diode body indicating the cathode (K) as shown in the picture below. Follow the mark on the diode for the polarity when connecting a diode.

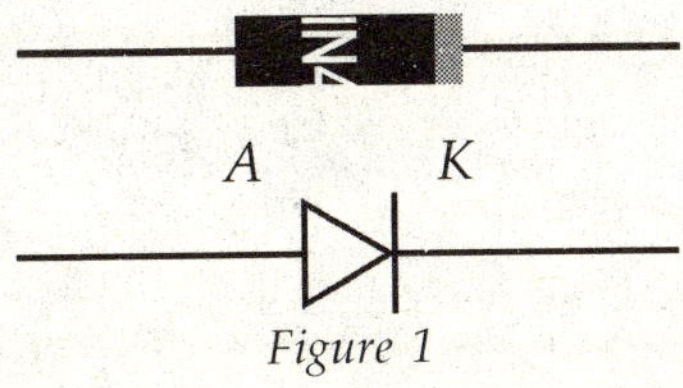

Figure 1

A word about capacitors

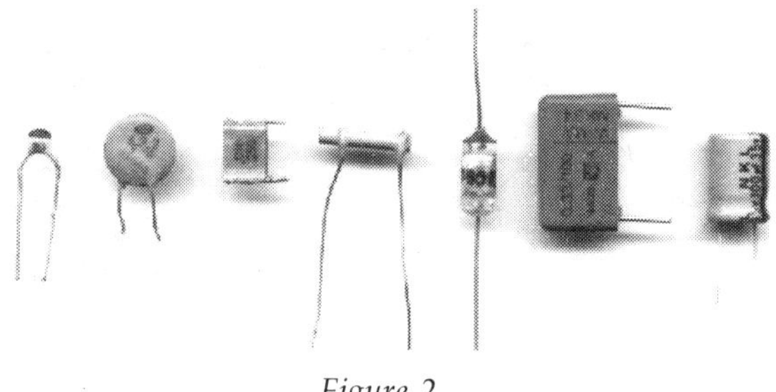

Figure 2

Capacitors store a bit of electrical energy and come in many types and varieties. Capacity is measured in terms of Farads named after the great scientist Michael Faraday. As farad is too large a value for most general applications, microfarads (mfd) and nanofarads (nf) are normally used in hobby and consumer applications.

Ceramic, polyester, paper, mica, and electrolytic capacitors are some of the types. A detailed discussion is left out here. Each type has distinctive features and qualities and capacities. Capacitors are marked with voltages at the maximum voltage they can operate. It is a good practice to use a capacitor rated at twice the operating the voltage. Capacitors above the range of 1mfd generally are electrolytics and are used such in these circuits. Electrolytic capacitors are polarized and marked on the can. They should be connected as per the polarity. Electrolytic capacitors of 25 V rating are used in the following circuits unless otherwise specified. Capacitors rated at higher voltages can be used in any of the circuits. Capacitors with higher capacity should not be used unless the implications are understood. Other capacitors normally rated at 30V are used in the following circuits unless otherwise specified.

A word about resistors

Resistors are measured in ohms, and have a wattage rating. Resistance value is marked on the resistor in an internationally accepted color code. Normally quarter-watt resistors are used in the following circuits unless otherwise specified. Carbon film and metal film resistors are generally used. Wire wound resistors are used for higher wattage resistors. Detailed discussion is left out here.

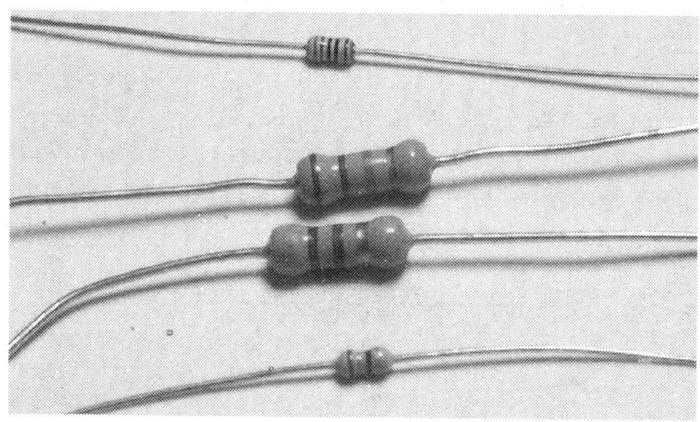

Figure 3

It will be useful to know a bit of batteries

Presently we have a variety of cells or batteries to choose from; lead acid, zinc carbon, Nickel cadmium and Nickel metal hydride, lithium, silver oxide and what not!

Cell is a storage form of electrical energy. When more cells are used, it is known as battery, but these words are used interchangeably. These are used where portability of the equipment is needed and some times as a back up for the mains voltage. Each these cells have their own characteristics, like cell voltage, current capability, storage capacity and rechargeability.

Good old Zinc carbon cells have 1.5 volts and if two cells are fixed in series, they show 3V. These are the most common cells powering most of the portable devices and are available in various sizes known variously as AAA, AA, BB, CC etc. This nomenclature denotes the capacity and dimensions of each cell but all these cells show a voltage of 1.5 volts. These are not normally rechargeable. Alkaline batteries last longer.

Lead acid cells are most widely used for high power applications such as in automobiles, UPS systems, emergency lighting systems where weight of the battery is no concern but the capacity required is high. Sealed lead acid batteries or maintenance free batteries are another form of the same where the electrolyte is within the moistened separators. Lead acid batteries can deliver high currents and recharge cycles are high. Each cell voltage is 2 volts and in a typical car battery, six of such cells are connected in series in a single container to give 12 V.

Nickel cadmium cells (Ni. Cad) show a voltage of 1.2 volts, which means that two cells in series show a voltage of 2.4 volts. These are rechargeable. They suffer from a memory effect. If they are recharged after only partially discharged, they tend to recharge only to that partial extent rather than to their full capacity.

Nickel metal hydride cells (Ni. Mh) also show a voltage of 1.2 volts and are rechargeable. They show certain better characteristics than nickel cadmium cells. They are environmentally friendly and do not show the memory effect of Ni-Cad cells.

Lithium ion cells are the fast emerging technologies. These have higher energy density and most of the present day cell phones, laptops; handy-cams are powered by these cells. They should be always used with protection circuit. Charge them only with specified charger. They have a cell voltage of 3.6 volts.

A word about veroboards

A general hobbyist does not have the capacity or inclination to make PC boards, hence we have the Veroboards. Veroboards are printed circuit boards with parallel copper tracks. Veroboards are used where ready-made boards are not available and after all you may not get ready-made PC boards for most of the hobby circuits. Components are located on these boards. Wherever not necessary, tracks are cut and where necessary jumpers are made.

Veroboards are also sold with copper dots with holes on a suitable grid. Botheration of cutting a track is avoided in these boards, but the copper on these boards is extremely delicate. A bit of heavy soldering will surely lift the dot from the board.

Take a Veroboard with holes matching IC pins as they are now available in different grids for holes and different patterns. Plan well how best the components should be inserted and cut the tracks where they are not required. Please make sure to cut the track fully as even a small copper left over can make enough damage. Join the component leads with jumpers between the tracks as required. Similarly it is better to make jumpers on the component side of the PCB rather than the trackside.

Figure 4

Breadboards

Breadboards are used to make trial and temporary circuits for testing or to try out an idea. They have sockets (called 'holes'), wherein the leads of most components can be pushed straight. As the holes are arranged on a 0.1" grid, ICs can also be inserted. Soldering is not required. Interconnections can be made with single-core wire of 0.6mm diameter. Parts can be changed and reused. Here we have a photograph of typical breadboard.

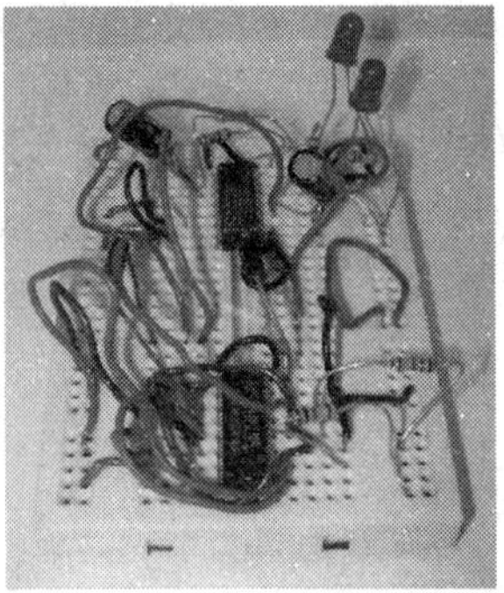

Figure 5

Transistor

Transistor is basically a three terminal device, used as switch or amplifier. Mainly there are two types of transistors, bipolar junction transistor (BJT) and Field effect transistor (FET). The circuits in this book use only BJTs. The three terminals here are called emitter, base and collector, labeled ***base*** *(B),* ***collector*** *(C) and* ***emitter*** *(E). Any one of these terminals can be made common between the input and output to configure a circuit, but most common is common emitter. Emitter and collector conduct* *current, when a small voltage, called forward bias, is applied at the base. There are two types of transistors, NPN and PNP, symbols of which are given below. These are made out of germanium and silicon. Presently silicon transistors are widely used. Here you can see various types of transistors. Have a look at the picture of the very first transistor invented by Shockley, Brattain and Bardeen which revolutionized twentieth century. We have come a long way.*

First transistor

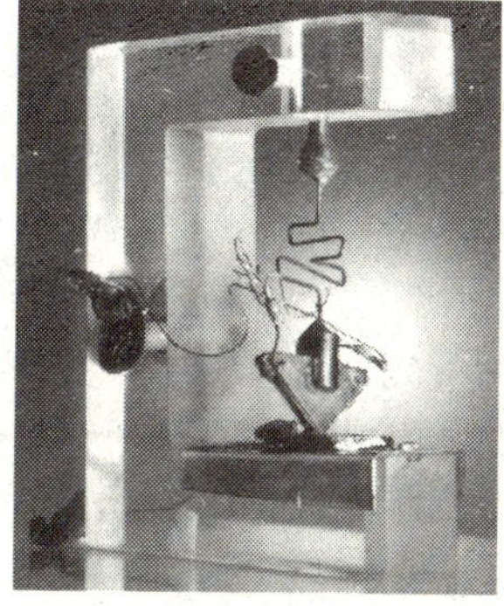

Figure 6

Some types of transistors now

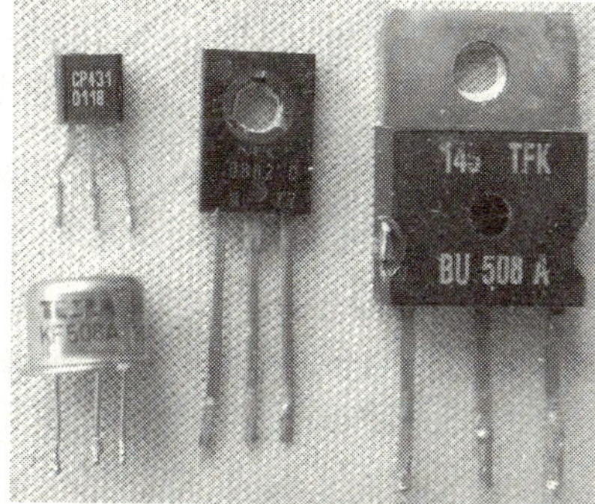

Figure 7

Transistor symbols

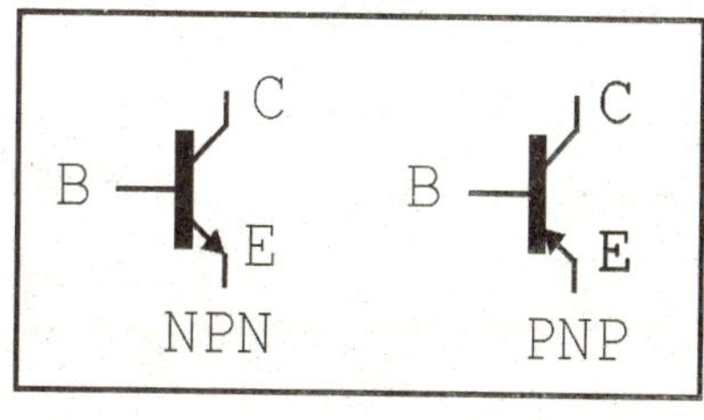

Figure 8

Integrated circuit (IC)

Integrated circuits actually consist of a number of active electronic components such as transistors and diodes along with other passive components such as resistors and capacitors interconnected as a single unit on semi conducting wafer usually silicon. Integrated circuits are classified as analog, digital and mixed signal (both analog and digital on the same chip).

Please find photograph (Figure 9) of the very original integrated circuits made by Jack Kilby. It contained about 4 transistors, whereas present microchips contain a few billion transistors. Picture is shown in Figure 10.

IC made by Jack Kilby

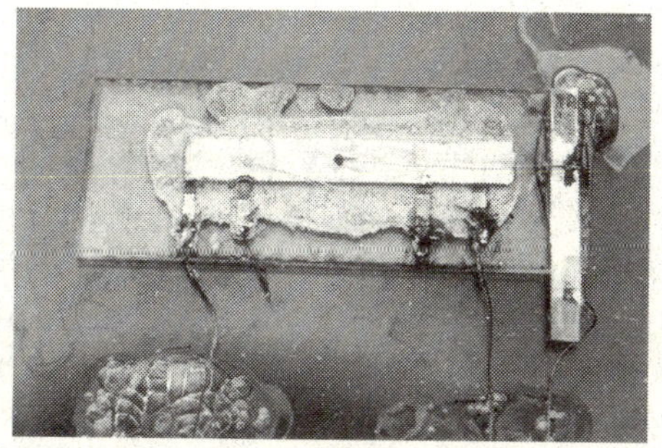

Figure 9

Figure 10

Regulated Power Supplies

Introduction

Unfortunately our hobby circuits work on different voltages and a power supply with single voltage will not be sufficient. Simplest way would be to buy a multi-tapped transformer and connect the voltage taps with a suitable switch. These transformers are available with taps of 1.5, 3, 4.5, 6, 9, 12V. We need a single multi throw switch (as many throws as there are taps). With 9-6-0-6-9 transformer and with a double pole double throw switch, you can get 6 and 9V DC. With the same transformer, and with a double pole six throw switch, it is possible to get 3, 6, 9, 12, 15, 18 V. Switches and multi-tapped transformers are difficult to procure; cumbersome and even unavailable at some places.

Such schematics are not shown here as there are better ways to build circuits to get variable voltages. Also these power supplies suffer from a major draw back; poor voltage regulation, the voltage falls as more and more current is drawn and voltage also changes with mains voltage fluctuations.

A 12 V power supply shown in **Schematic 3** may show as much as 16.8 V without load and it may go down to less than 12 V as the load current is increased. This still can be and is safely used for most of the non-critical circuits, but it is an unstabilised supply.

Why not build a regulated power supply, which doesn't cost an earth but still gives very good results. Now for the time being, without going into details and complications of integrated circuits, we will use them for the pleasure and ease of using them and build a regulated power supply as a starter.

We have some wonderful devices in three terminal regulators, just three terminals, IN, GROUND, OUT. How's that?

Build one!

Three Terminal Regulators

Description

LM7805, 7806, 7808, 7809, 7812, 7815, 7818, 7824.

These are simple, effective and interesting integrated circuits. They are short circuit protected, can stand up to 32V, have thermal protection and they can easily deliver regulated voltage with 1 amp of current. In these ICs, the last two digits give you the regulated voltage they deliver, 7806 gives you 6 volts and 7812 gives you 12 volts and so on. These are available from most of the manufacturers and are cheap. They are connected after the conventional power supply. There should be 3V more than the required voltage at the input for proper regulation. You need 9V for a 7806 regulator. Use this circuit whenever single regulated voltage is required. Pinout and heat sink mounting procedure are given in figure 11 and Figure 12 respectively.

Construction

You have already built the bridge, i.e., DC power supply with a bridge rectifier. Just solder the three terminal regulator after the bridge. Connect the input pin (1) after the bridge and capacitor. Connect the common terminal (2) to the negative rail and you have the regulated voltage at the output (3). 47 uF capacitor (C2) is required if the regulator is placed faraway from the main power supply.

Metal bodies of these ICs are connected to the ground (common) terminal. Follow the mounting procedure as given below and fix a small aluminum plate to remove the heat and to improve its current capability. (This is a standard practice to mount heat sinks on ICs, transistors, SCRs, and triacs, etc.) By now you must have caught up with the art of soldering and the IC should be soldered well and fast.

Schematic 4 shows a 9V regulator appended to 12V power supply of the earlier bridge power supply of 12V DC.

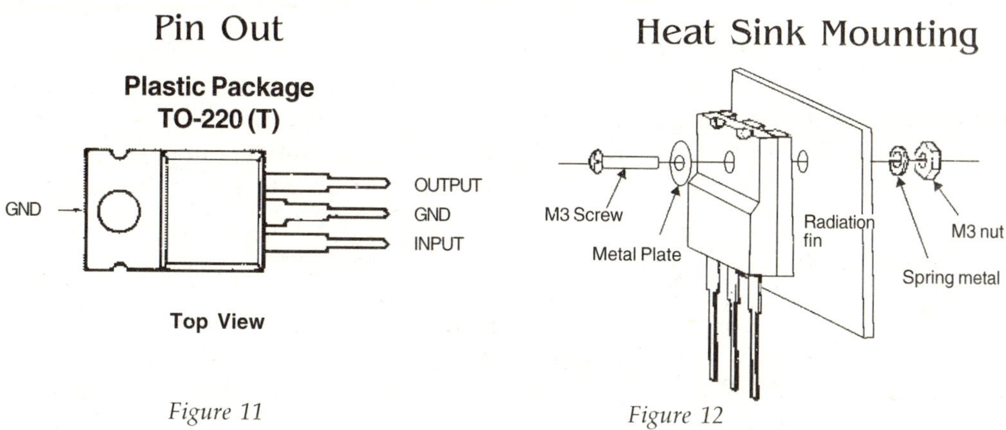

Figure 11

Figure 12

Parts

Additional Parts required over bridge circuit are 7809 and 47uF capacitor.

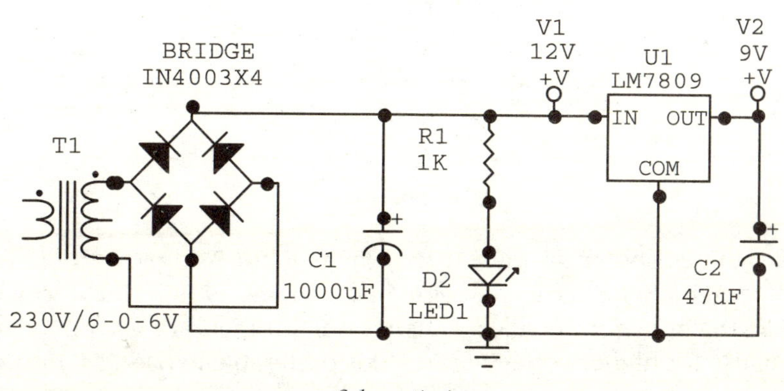

Schematic 4

Three Terminal Adjustable Regulator

Introduction

Three terminal regulators are just right when we need a single fixed voltage. But they will be cumbersome as a hobbyist works with different voltages at different times and it becomes unwieldy to keep power supplies with a number of three terminal regulators!

Wouldn't it be better if we have an adjustable but regulated power supply, which can cater to all the voltages we need? And we have such a wonderful device where you can continually adjust your output DC voltage. Well! It is short circuit protected, has only three terminals and all other good things. LM317!

Description

LM317T is an adjustable 3 terminal positive voltage regulator capable of supplying around 1.5 amps over an output range of 1.25 to 37 volts. It also has built in current limiting and thermal shutdown features, which makes it virtually blowout proof. This is an excellent startup project with low ripple. With an easy adjustment of regulated voltage, it can be used as power source for most of the applications in the next chapters. Pin out is given in *Figure 13*.

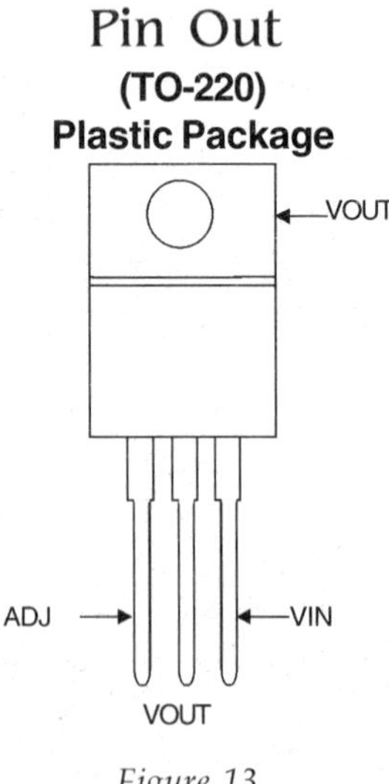

Figure 13

Rectifier part is same as shown in the earlier schematic of bridge rectifier. Transformer is changed to 18 V to get a better range of regulated voltages. 9-0-9 Transformer is used as it is difficult to get an 18 V transformer. Please change transformer rating as per your requirement and the capacitor also for higher voltage rating. It is preferable to use 2200 mfd. caps as larger value makes good, low ripple output voltage.

Pulsating DC output from the bridge is now filtered by the 2200μF capacitor and fed to 'IN'-put terminal (1) of LM317 regulator. The output of this regulator is varied via the 'Adj' pin(3) and the 5K variable resistance or preset pot meter (R1) connected to it. The regulator uses an internal Zener diode to provide a fixed reference voltage of 1.2 volt across the external resistor R2. Hence the lower end of output voltage is limited to 1.2 volts. C2 is 47μF decoupling capacitor to filter out the transient noise. Metal tab of LM317 is connected internally to the 'Output' pin (2). The circuit diagram is shown in Schematic 5.

Construction

Use a small Vero board to fix all the components, cut the tracks where not required and solder the pins. Mount the LM317 regulator on a heat sink. However it can be screwed to the metal case of the enclosure box with the mica insulator and the nylon washer with the mounting screw as shown in ***Figure 12.***

Use a little of heat sink compound on the metal tab and mica insulator as it helps to transfer heat between LM317 and case or heat sink.

Use a metal box of suitable size and fix the veroboard and transformer. Mains wire is connected to the primary side of the transformer and taken out. Carefully insulate the mains joints. Mains switch is not shown in the schematic. You may add one if it is required. 5k linear potentiometer is fixed at the casing. Measure the voltages and mark them suitably on a dial fixed on the face of the casing with appropriate indication of the voltage. LED is also fixed on the casing to indicate the power supply.

Parts

Item	*No. reqd*	*Description*	*Designation*
1	1	5k variable	R1
2	1	240 ohms	R2
3	1	1.5k	R3
4	1	47 uF	C2
5	1	2200 uF	C1
6	1	LED	D1
7	4	IN4003 DIODES	D2, D3, D4, D5
8	1	9-0-9 TRANSFORMER	T1
9	1	LM317	U1

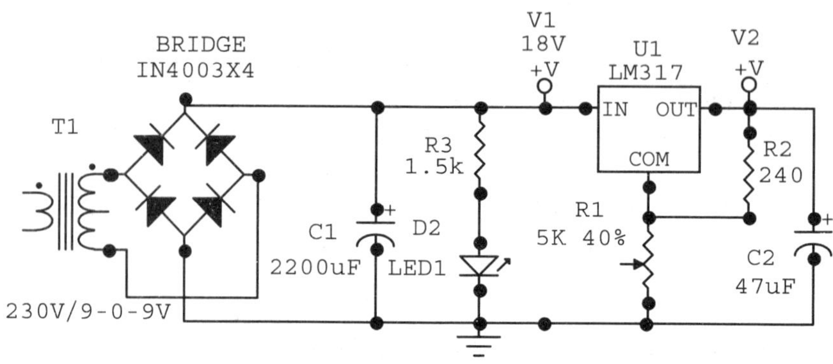

Schematic 5

With power supplies built, jump to any of the following sections and build a project of your choice. What do you want to build? Before we go and build a project with an integrated circuit, why not we try our hand on a simple device and learn a few skills. These are the projects for starters and if you have certain skills in soldering and electronics, you may safely skip pages.

Buzzers

Introduction

These are simple projects and yet they are very useful. You need not have the soldering skills to work on the following piezo buzzer projects. Often the terminal strips used by the electricians will suffice. We can build excellent domestic alarms with these delightful devices known as piezo buzzers. Piezo devices are everywhere; gas lighter in the kitchen for instance. Piezo alarms are seen nowadays, in cars as reversing tunes and turning indicators, alarms in watches, and beepers in washing machines, ringers in telephones, and musical tones in electronic games, indicators in calculators, computers, and ATMs. Piezo transducers are also used as pressure and strain sensors and position sensors.

Figure 14

These buzzers are available with 2 terminal or 3 terminal models. When activated, these devices give out a fairly loud sound up to 85 dB, in continuous, intermittent, or in musical notes. These rugged devices can operate generally from 3 to 26 V and will consume about 10 milliamperes of current. But to get full volume out of the device, do not close the sound emitting hole or keep any obstruction within 15 mm of the device.

Three terminal buzzers are used for the following projects. These simple but extremely useful and effective buzzers have three wires coming out, appropriately marked generally with colors, red, black and yellow respectively for positive, negative and trigger terminals. We have water high level alarm, water low level alarm, and fire alarm, which can of course be easily modified to continuity tester and steady hand game. Half wave power supply would suffice for these applications.

Note: If you are well-versed with buzzers, skip to the next section

Water High Level Alarm

Introduction

This can be used as an alarm when filling overhead tanks. Quite some time ago, I used it as an indication when the municipal water used to start flowing to my pipeline. When I built this project, happiest one was my mother. Municipal water used to be pumped in the midnight at unearthly hours and my mother used to wake up now and then to check if the water started pouring in. Until I built this circuit! The circuit is shown **Schematic 6**. It as simple as that!! The project will cost less than Rs.50/-

Description

With half wave power supply already built, connect the buzzer as shown in the schematic and fix the wires. As the water level rises to touch the wires, the buzzer gives out a loud alarm. I used to put these wires in a plastic bucket under the tap. When municipal water flows down, it indicates an alarm. It is one of those most useful and easiest projects I ever built.

You may use this as a rain alarm particularly if the lady of the house dries up clothes or grain in the open yard or even as bed wetting alarm for the baby ((Use only battery power!!)).You can try this as a steady hand testing game. You may use this for a continuity tester but do not use this on live wires.

Construction

Bring two wires and keep them at the desired level in the tank with a little separating distance between them. You can use fairly long wires and run them along into the house. Needless to say scrap and expose the copper wires a little. Connect one of these wires to the trigger terminal (generally yellow) of the buzzer and the other to the positive terminal (red wire). Black wire goes to the negative.

If you want battery operation, use two battery cells (1.5V each). Use more cells for more sound. If you want to use mains supply, half wave circuit description is given above. Use a transformer with 12V or 6-0-6V at the secondary. Screw the secondary ends to the terminal strip; screw diode (D1, IN 4003) C1 is an electrolytic capacitor of 1000 mfd and 16V rating. Please follow the polarity marked on the can. Light emitting diode is optional. Use with resistor (R1, 1K). It gives an indication that the power supply is OK.

You may use the mains plug mounted power supplies or eliminators and mount all the parts on it.

Parts

Item	*No. reqd*	*Description*	*Designation*
1	1	12V Transformer	T2
2	1	IN 4003 Diode	D1
3	1	1000 uF /16 V Capacitor	C1
4	1	1K Resistor	R1
5	1	Light Emitting Diode	D2
6	1	Piezo ceramic Buzzer	BZ1

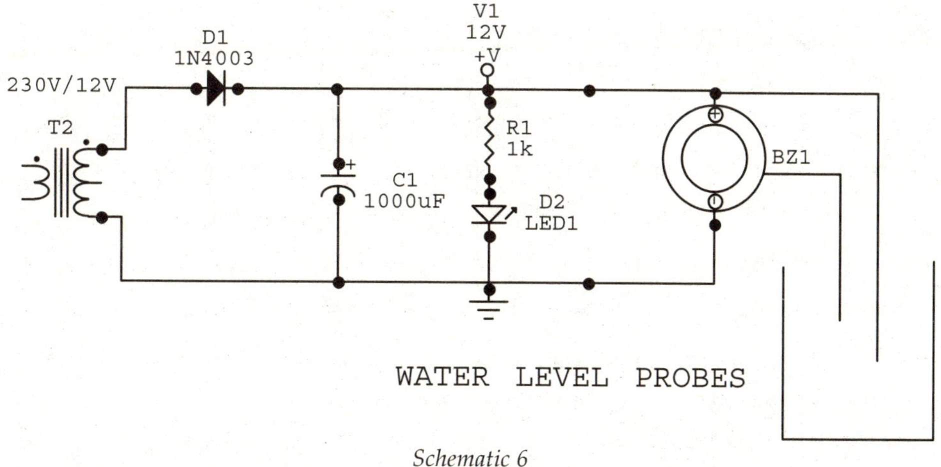

Schematic 6

Water Low Level Alarm

Introduction

Same buzzer, the same power supply but with a little ingenuity this gives an alarm when the water falls below a certain level.

Description

This circuit as shown in **Schematic 7,** which can be used to indicate when the level falls low in overhead tank, a sump or even in a bore well.

Now you will find a 100K resistance between the trigger terminal and positive supply which pulls the trigger terminal to the positive potential. As in the previous case both wires from trigger and ground are kept in the water. Even though pulled up, trigger will be at ground potential as long as there is water across the probes and hence the piezo buzzer can not go off. Once the water level goes down, the trigger terminal loses its ground potential and reaches the positive potential because of the pull up resistor and the buzzer will be on.

Digital Circuits

Incidentally in digital electronics parlance this is known as going high or pulling up. If the voltage at any terminal is close to the ground potential as dictated by system requirement, it is known as going low or pulling down. Conversely if the voltage at any terminal is close to the positive potential as dictated by system requirement, it is known as going high or pulling up. Digital circuits work usually with 0 and 1. Obviously one is high level and zero is low level, which are known as logic states. Noise in the digital circuits can easily be eliminated. Follow the logic states depicted here.

Logic states	
True	***False***
1	0
High	Low
+Vs	0V
On	Off

Figure 15

Analogue Circuits

On the other hand analogue circuits process analogue signals which can output any value within a range of its power supply. Examples are audio, video amplifiers. Unlike digital circuits, noise here is difficult to control.

Construction

Construction is similar to earlier project, except that the probes are now fixed at the required low level of water.

Parts

No additional parts are required except 100k resistor-1/4 watt.

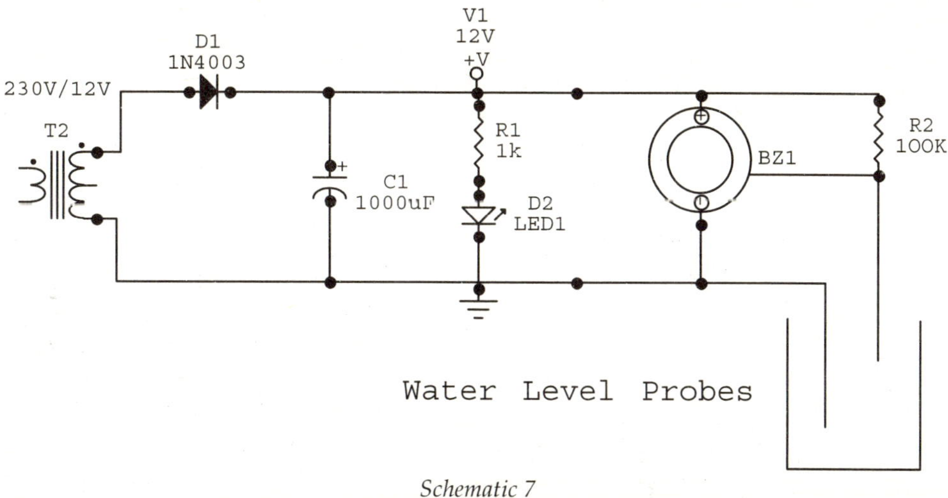

Schematic 7

Fire Alarm

Introduction

Just a little more ingenuity! Just by adding one more component, here we have an application for a fire alarm. This can be used in cellars, stores and unmanned places to indicate if there is fire in the place.

Description

The circuit makes use of characteristic of a diode. As already explained normally diode conducts only in one direction and it blocks voltage in the reverse direction. However there will be a little leakage current flowing through which is normally negligible. But if the temperature increases, this leakage current also increases and this effect is used here to create a fire alarm. This circuit as shown in **Schematic 8.**

Construction

Use any IN 4000 series diode as a fire sensor and connect the rest of the circuit also as shown in the diagram. Do not cut the leads of the diode short. It is better to use the full length of the leads as far as possible. For a trial, bring the tip of your hot soldering iron and touch the terminal wire of the diode and you hear the sound soon enough. Or bring a lighted matchstick. You will notice that the sound increases as the heat increases. If the diode is fixed the other way round, (or in other words, if it is reverse biased) the buzzer continues to sound, fire or no fire.

Ah! That's a tutorial on reverse biasing.

Parts

Additional part required is only another IN 4003 diode.

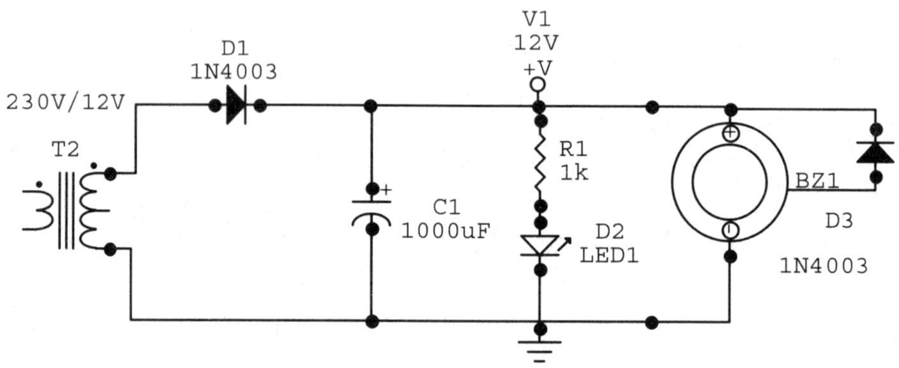

Schematic 8

There is nothing more thrilling than building an alarm and enjoying the response. Electronics is exciting. In the following pages you find a lot of alarms but before we build any of them, let us tackle one of the most common ICs. If you worked on 555, you may safely skip to the next section. But if you want some theory and an insight into it, go on.

IC 555

Introduction

IC 555 has been one of the most popular ICs since Signetics introduced it in the early 70s. 555 is also called as a TIMER IC. This 8-pin chip is a versatile device suitable for astable, monostable and bi-stable operations. It is such a multipurpose device that hobbyists find clever uses of it everyday. It is so rugged that it is safe and thrilling for making a good beginning.

Before we start, let us learn just a word about astable, monostable, and bistable without again going into technical details.

Astable is an oscillator. In other words it generates a continuous wave.

Monostable operation is also called one shot and it is basically a time delay.

Bistable operation is one in which the output goes either high or low.

Internally IC 555 contains more than 28 transistors. We only need to fix two or three external components to produce an oscillator with a frequency from 1Hz to 500 KHz or obtain a time delay from a few microseconds to hours. It can directly drive a small speaker for audio alarms delivering about 250mW of power.

Here we have the pin out in *Figure 16* and a small description of the pin assignments, without the bothering details of upper comparator, lower comparator, threshold, control flip-flop, and discharge, etc.

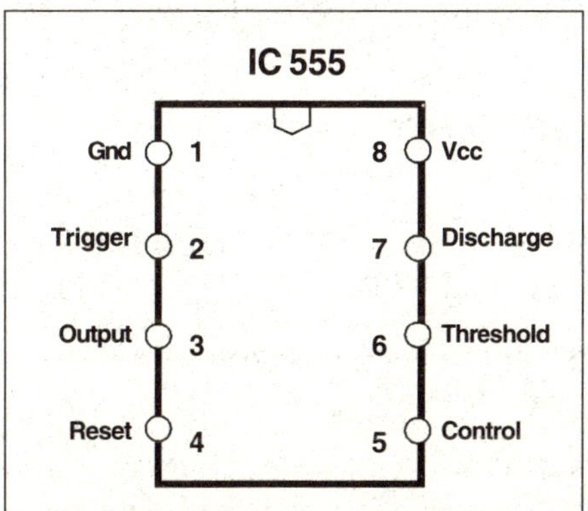

Figure 16

Pin 1 is the ground pin and connected to the negative rail.

Note: **If you are well-versed with 555, skip to the next section.**

Pin 2 is the trigger terminal.
When this terminal is made low, it causes the output high. Further input of low level during the mono-stable operation will not have any effect. This pin prevents the chip from starting the timing cycle as long as it is high.

Pin 3 is the output pin.
It can source or sink currents up to 200mA. 555 must be used with a driver for currents higher than 200mA. Momentarily taking the trigger pin (2) low makes the output high. Output can be made low by pulling the reset pin (4) down to low or by lifting the threshold pin (6) from low to high.

Pin 4 is the reset pin.
It overrides all other functions. It is normally tied to the positive supply for normal operation. When made low, it pulls the output down. Contrary to the pin 2, this pin stops the chip from working when it is taken LOW.

Pin 5 is the control voltage pin.
By applying a voltage to this pin, the timing of the chip can be varied to a certain extent independent of the RC network, which means that the output can be modulated. In other words, pin 5 can adjust the waveform. This pin should be bypassed with a10kpf capacitor to prevent noise into the chip.

Pin 6 is the threshold pin.
It is an input pin of the upper comparator. When this pin is made more than 2/3 of the supply voltage, the output of the IC goes low.

Pin 7 is the discharge terminal.
The timing capacitor is connected between this pin and ground. By design, any value of capacitor can be connected without the worry of charging currents.

Pin 8 is the supply voltage terminal.
Any voltage between 4 to 15 volts can be used for 555 and IC will operate without significant change in timing.

The formula for time in monostable operation is

$$T = 1.1 \times R \times C \text{ (in seconds)}$$

The formula for frequency in astable operation (oscillator) is

$$f = 1/\ (.693 \times C \times (R1 + 2 \times R2))$$

Let us build some projects with 555. 12V half wave power supply will do well for these projects. Full wave is better naturally.

IC555 is versatile and can be used ingeniously for a variety of applications. 7555 is available as a CMOS version, pin to pin compatible with normal 555. 7555 can be used in all these circuits. It is faster, can handle higher frequencies and has lesser current consumption. But whereas 555 can source or sink 250mA, 7555 has an output current of 50mA only. This is important when 7555 IC is driving low impedance loads such as speakers.

Sinking and Sourcing Current

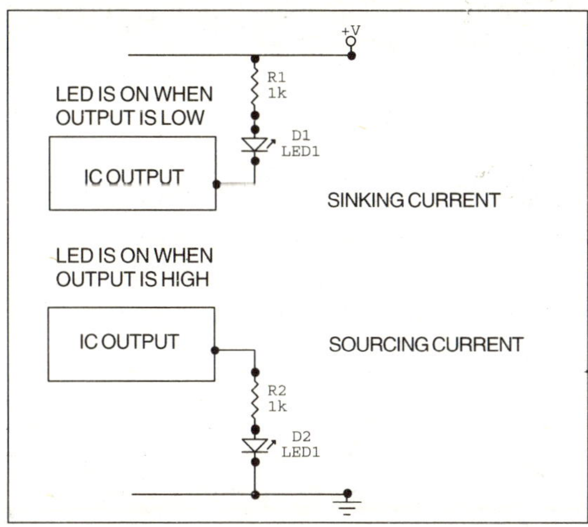

Figure 17

You must have observed that while discussing pin 3 of 555, we talked of sinking and sourcing. Chip outputs are often said to 'sink' or 'source' current. The terms refer to the direction of the current at the chip's output. If the chip is sinking current it means that the chip output will be switched on when the output is low (0V). If the chip is sourcing current means that chip output will be switched on when the output is high (+Vs). It is possible to connect two devices to a chip output so that one is on when the output is low and the other is on when the output is high. This arrangement is used in projects to make LEDs flash alternately. 555 can do it. Have a look at ***Figure 17.***

Basic Oscillator (Tone Generator) At 1.8 KHz

Introduction

Here we have an astable oscillator built around 555, which gives an alarm tone of 1.8 KHz directly driving a speaker. This is basic alarm circuit, which is later used in many other projects in the book. While the circuit shows only 1.8 KHz operation, it can be adapted to different applications by changing frequency and duty cycle after manipulating the timing components as per the formula.

Description

The circuit is shown in **Schematic 9.** Resistances R1, R2, and capacitor C determine the frequency. Changing these values change the frequency. You can also manipulate the tone by changing the capacitor at Pin 5. You can modulate the tone also. Try connecting a 10 K resistor from secondary of the transformer, i.e., before the rectifier and connect it to Pin 5, the Control terminal. Output will wobble at the mains frequency of 50Hz. Output at Pin 3 is coupled to the speaker of 8 ohms. The capacitor C2 is essential to block DC. You can also vary the frequency by adding a preset potentiometer in the in the timing components. Try changing the timing capacitor also. You can increase the volume by amplifying the output with a transistor. You will find such applications later.

Construction

Construction is straightforward. Use a small Vero board suitable for ICs. Mount the capacitors and resistances close to the board. C2 is an electrolytic capacitor. IC should be fixed finally and if you have not got the skill of soldering well, use of IC base will be helpful. If you wish to vary the tone, solder 100K variable resistance (preset), between power supply and 1K (R1) and change the tone as you like it.

Parts

Item	No. reqd	Description	Designation
1	1	.1uF	C1
2	1	250uF	C2
3	1	.01uF	C3
4	1	1K	R1
5	1	3.3K	R2
6	1	8 Ohms	SPK1
7	1	555	U1
8	1	Piece of veroboard	
9	1	100K preset	Optional

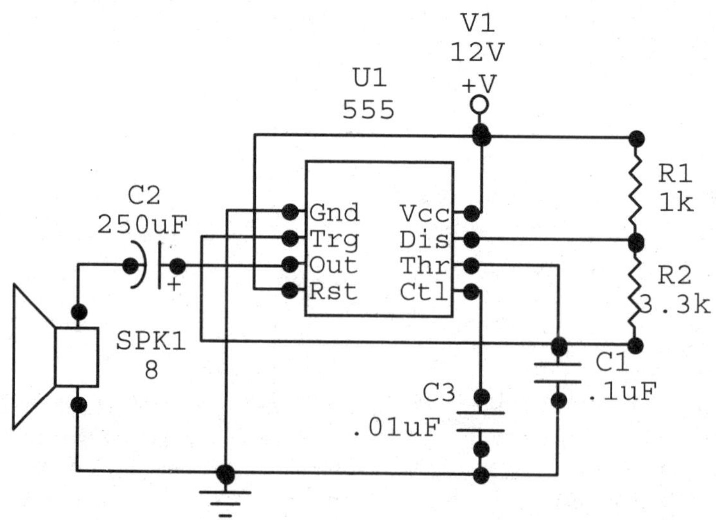

Schematic 9

Water High Level Alarm

Introduction

This is a big brother of the earlier piezo version, which can be used as an alarm when filling overhead tanks or as a water tap flow indication. Unlike the piezo version, this alarm can give out fairly loud tone and can even be amplified further up. Tone can also be varied to suit individual needs.

Description

The circuit is shown in **Schematic 10.** You would have noticed in the earlier circuit that the trigger terminal (2) is connected to the threshold terminal (6) to make it an astable oscillator. Now the trick here is that these terminals are coupled through water.

Construction

There is not much of a change from the earlier circuit. Remove the jumper connecting Pin 2 and Pin 6 and connect two wires from each terminal and use them as probes. These wires can be any hook up wires or wires used for the domestic wiring. Scrap the ends of the wires a little and use them as probes. Pure water is non-conducting but domestic water and water supplied from the wells conducts a little bit and that is enough for this alarm to operate. You may have to occasionally clean these ends, when you notice scaling.

Parts

Parts are same as the above circuit. Only you need long wires.

Note

This alarm can be modified to check out if the overhead tank is empty. This is more important as the tanks have inevitable tendency to get empty at a wrong time just as the guests arrive and the power also goes out just about the time. This can be used to forewarn impending low level. Such an alarm can be used to check if the sump level is low, or if the water level in the cooler has gone down. The circuit can be adopted to give an alarm when the water level goes low by manipulating wiring at 6 (Threshold) and 2 (Trigger) pins or by the pin connections at 4 (reset) and 1 (ground).

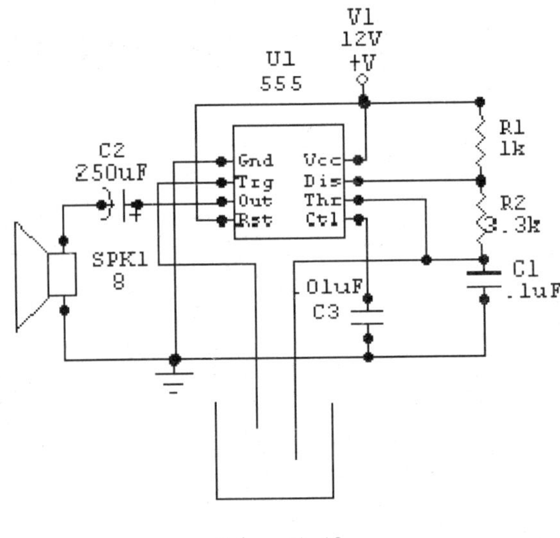

Schematic 10

Light Alarm (Sun Up Alarm)

Introduction

This circuit designed to sound a loud alarm at the break of the dawn especially for those who cannot wake up even with an alarm clock. Circuit can be modified as an intruder alarm or as a security alarm. Needless to say that 1.8 KHz astable multivibrator with 555 is used here again with a light dependent resistor used as a light sensor.

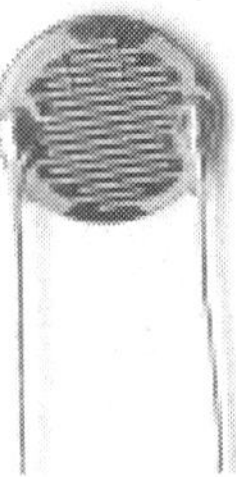

Figure 18

Light dependent resistor is made of Cadmium Sulphide and it offers very high resistance (about 1 mega ohm) when there is no light falling on it. If the light falls on it, resistance falls sharply (about 100 ohms). LDRs are popularly used in light sensor circuits. They are cheap and reliable and we can build a number of light sensing circuits around LDR

Description

The circuit is shown in **Schematic 11.** When there is no light falling on the LDR (R5), the transistor (Q1) is cut off as it is reverse biased by the variable resistor. Hence the reset pin of the 555 is pulled low and the 555 stays reset. Resistance of LDR decreases as light falls on it and the base of the transistor becomes forward biased. Q1 turns ON and pulls the reset pin 4 of the 555 high. This starts 555 as an oscillator and the sound produced at Pin 3 (OUTPUT) is further amplified by Q2, driving the speaker. Q2 is added here to increase the volume. You can still use this without amplification as in the earlier circuits. Variable resistor 100K (R3) is the sensitivity control. 10K resistor (R7) in series with protects the transistor Q1 when the preset is set to zero resistance. Do not remove it.

You will find now that BC 157 transistor works as a switch and BEL 187 as an amplifier. BC 157 is a PNP transistor and BEL 187 is an NPN transistor.

Pin out of BC147, BC157 and BEL 187 transistors

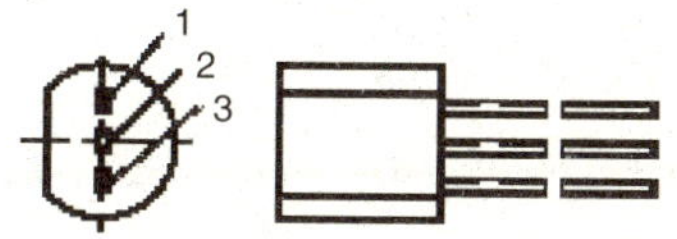

1. Emitter, 2. Base, 3. Collector

Figure 19

Construction

If you have already built the earlier oscillator, construction of this should pose no problem. We have only added two transistors, a LDR and a few resistances. LDR should be conveniently placed near the outside light and away from room light or any other light to avoid false triggering. Use a Vero board. It is very important to follow the pin out of the transistors. As far as possible do not cut the leads short. Use sleeves on individual leads so that they do not short between each other.

Note

The circuit can be modified to give alarm when the light goes out. Keep the transistor Q1 and its connections intact along with LDR and R7.

LDR- Remove the end connected to ground and connect it now to power supply line.

R7- Remove end connected to power supply line and connect it now to ground.

Now by directing a light source straight into LDR, the alarm will be off as long as there is light falling on the LDR and anybody crosses and breaks the light beam, alarm will be on. This way the circuit can be modified as an intruder alarm or as a security alarm.

Parts

Item	No. reqd	Description	Designation
1	1	.1uF	C1
2	1	.01uF	C3
3	1	BC157	Q1
4	1	BEL 187	Q2
5	2	1K	R1,
6	2	3.3K	R2, R6
7	1	100K	R3
8	1	4.7K	R4
9	1	8 ohms	SPK1
10	1	555	U1
11	1	LDR	R5

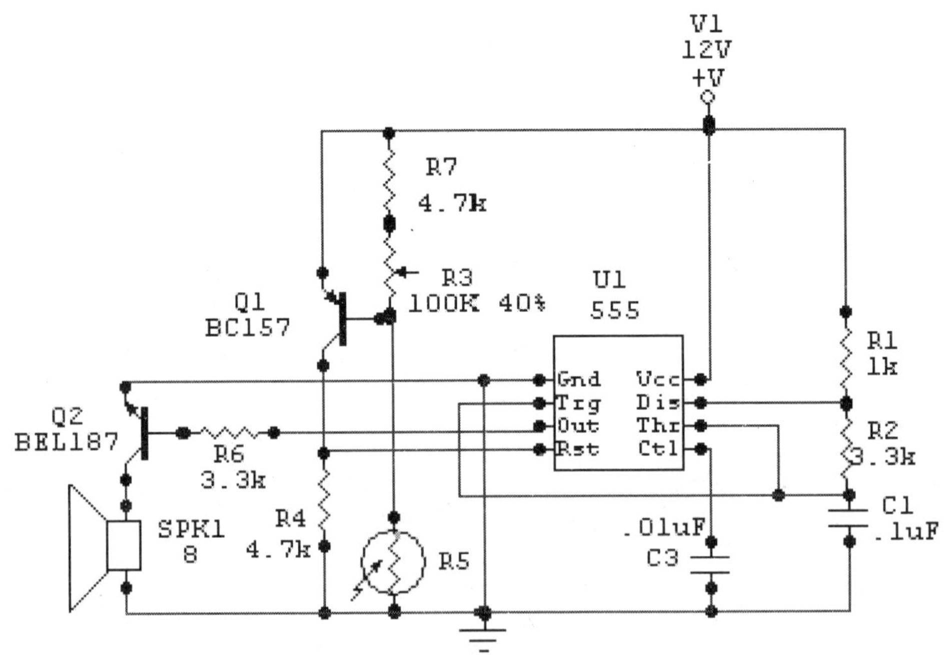

Schematic 11

Burglar Alarm for Car Stereos

Introduction

We have another interesting application of 555. This is used as a burglar alarm for car stereos. Thieves have, some how, the knack of removing the car stereos even though all the car doors are closed and locked. If we forget to close any window, these people will have a field day. This circuit can easily be adopted as a domestic burglar alarm. The original astable oscillator of 1.8 KHz is still used but with some small tricks. As this alarm should alert passers by even on a busy road, more volume is developed by a Darlington pair. You may use this in the house as a burglar alarm or in the garden.

Description

The circuit is shown in **Schematic 12.** Car stereos are normally kept in metal shell, which is clamped into the housing meant for auto stereo decks in the car. Body of the car stereo is normally connected to the ground that is the negative side of the battery in the present day cars. Shell is also connected to the car in some models by a separate wire coming from the wiring harness or the terminal block of the stereo connections. That suits us well. The alarm is placed at a convenient place and power is taken from the battery before the ignition switch so that this alarm is always powered. Now the reset terminal is pulled up by R3 (10k resistance), but it is also connected to the ground from the shell of the stereo deck so that the connection is made only when the stereo is inside. Hence it is effectively at the ground potential as long as the stereo is inside the shell. If any body removes the deck from the shell, reset pin looses ground connection and goes high. Alarm sounds.

To give a very loud volume, a Darlington pair is used now comprising of Q1, Q2. *Darlington pair is a cascade of two transistors where the gain of individual transistors is typically multiplied. Collector load is shared by both transistors. It works as if single transistor of a very high gain is used.*

You may use this circuit for protection of home.

Use a very thin enamel wire (L1) of; say 36 gauge wire is connected between the reset pin and ground. This can be a long wire running invisibly across the door, window or across any other places you want to protect. You can even wire at around a fruit bearing tree in the garden. As the reset pin is shorted to the ground with this wire, oscillator will not start. If the wire is broken, reset is pulled up and the alarm starts.

Construction

The alarm in a suitable box is placed outside the stereo at a convenient place. Please see if the shell is already connected to ground or if it is making the ground contact from the casing of the stereo. The trick is to keep the reset pin at ground level during normal operation. If the contact is broken, alarm sounds. If the stereo casing is connected to ground, take the ground contact of reset pin from the shell or vice versa. If they are totally housed in plastics, use a very thin enamel wire and connect the shell and the casing. Connect one end to reset pin. Negative or ground connection is common with the automobile ground. Now if the stereo is removed from its position, ground contact of the reset pin is broken. It goes high and alarm sounds.

If you are using this around the house, use a length of thin enamel wire. Fix it around the area as required. Scrap the enamel and solder the ends at Pin 4 and ground. Handle this wire carefully as it is likely to break while working. Construction is same as the 1.8K astable oscillator. Reset Pin (4) is pulled up with a 10K resistor.

This project may have to be left on for considerable lengths of time. Hence it will be useful to use CMOS version of 555, i.e., 7555 for this project as its standby current consumption is very low. The Darlington pair amplifies the output any way. CMOS version can work at higher voltages.

Power is taken from the car itself. It should be taken before the ignition switch, directly from the battery as the circuit needs to be powered always.

Parts

Item	*No. reqd*	*Description*	*Designation*
1	1	.1uF	C1
2	1	.01uF	C3
3	1	WIRE LOOP	L1
4	1	BC148	Q1
5	1	BEL187	Q2
6	1	1K	R1
7	1	3.3K	R2
8	2	10K	R3, R4
9	1	8 Ohms	SPK1
10	1	555	U1

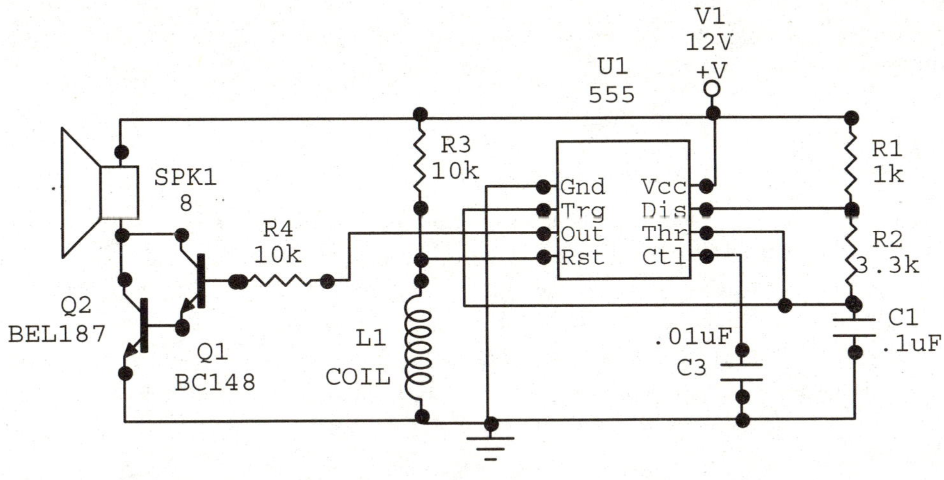

Schematic 12

Light Dependent Staircase Switch

Introduction

One of the earliest projects taught to the electrical engineers and technicians is a staircase switch, the technique of which I could never master any time. However here we have electronic application of the same loaded with added features.

In the conventional staircase application, the lights can be switched on only from two places, normally from a switch at top of stairs and another at the bottom. In this case there can be even more switches like from inside of a room also. As an added feature, if there is sufficient ambient light, the staircase lights won't come on at all even pressed from any switch. And in the conventional circuit two-way switches are used, but here simple push-to-on switches will suffice. Another feature is that the light will go off after three minutes after switching on. You can forget switching off at the top of stairs.

Description

IC 555 is wired as a monostable, i.e., a timer application. With the components shown at Pins 6 and 7 i.e., R1, R2, C1, the time delay is about 3 minutes, which should be adequate for normal use. Once the circuit is triggered by any one of the switches S1, or S2, Pin 2 is pulled down. 555 goes into timing operation and output goes high at Pin 3 driving Q2 with R6. Relay RLY1 is fired and D1 suppresses reverse voltage of the relay operation.

LDR R5 senses the ambient light, along with the sensitivity control R3. If there is sufficient light, Q_1 is not triggered. Consequently Reset pin 4 is pulled down by R4. In this condition

555 cannot work even if S1 or S2 is pressed. If the ambient light is not enough, Q_1 conducts and pulls reset pin 4 high. IC 555 will be ready for operation when switched on by S1 or S2. The circuit is shown in **Schematic 13.**

Construction

R3 (100k) is a preset potentiometer used as a sensitivity control, which can be adjusted for the level of light at which the circuit can be triggered. LDR should be placed in such a way that routine lighting disturbances would not false trigger the circuit. Timing can be increased or decreased by changing the values of R1, R2, C1 at Pins 6 and 7. At higher values of C1, it is good to use tantalum capacitors, as timing variations occur with aluminum capacitors. However accuracy of timing is not critical in this application. Relays should be rated at 12V and have 300 ohms of coil resistance. Contacts should be rated at least 1amp as about 100 W light bulbs are generally used.

Note

The circuit can be modified for direct mains operation. You may take a cue from the circuits that follow and make one suitable for placing in the wall socket.

Parts

Item	*No. reqd*	*Description*	*Designation*
1	1	47uF	C1
2	1	.01uF	C3
3	1	1N4003	D1
4	1	BC157	Q1
5	1	BEL187	Q2
6	5	10K	R1, R2, R6, R7, R8
7	1	100K variable	R3
8	1	4.7K	R4
9	1	LDR	R5
10	1	Relay 12V/300 Ohms	RLY1
11	2	Push to on switches	S1, S2
12	1	555	U1

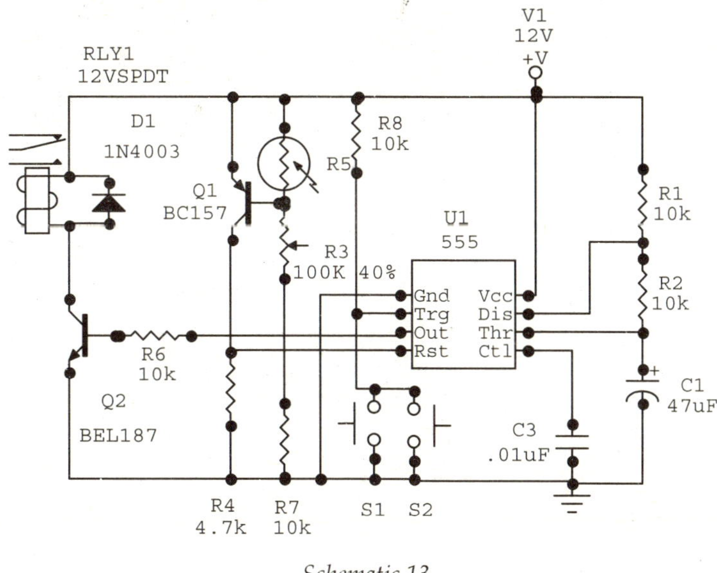

Schematic 13

Car Battery Low Indicator

Introduction

Almost every one of us has fallen prey to a flat battery at one time or the other. On the lighter side, batteries some how have an uncanny sense of failing at our urgent moment. Here is circuit that gives a general indication of the health of the car battery and warns you to give it a boost. Do not blame me for another 555, after all it is so versatile. One more automobile circuit, but a very useful one!

Description

The circuit is shown in **Schematic 14.** Here again the 1.8 K oscillator is used as the alarm. However it is disabled by transistor Q_1, whose base is driven by 10V Zener diode. As long as the battery voltage is above 10V, this diode conducts, transistor Q_1 also conducts. It clamps the reset pin of IC 555 to ground and the oscillator cannot function. When the battery voltage falls below 10V, the transistor cannot hold reset pin to ground and the oscillator starts functioning. 555 then gives an alarm indicating low battery voltage. LED D2 will light up as long as the voltage is good.

Zener diodes are used to regulate the voltages. When forward-biased, zener diodes behave like normal rectifying diodes, but in reverse-bias, they do not conduct until the applied voltage reaches or exceeds the so-called zener voltage. Then it conducts substantial current, and while doing so, it will try to limit the voltage to that zener voltage. As long as the power dissipated now does not exceed the diode's thermal limits, the diode will not be harmed. Zener diodes are widely used to regulate the voltage and they are manufactured from a few volts to hundreds of volts.

Construction

Construction is similar to other 555 circuits and needless to say th1at the power supply is from the car battery. The circuit will be placed in an automobile and has to be necessarily rugged. Automobile body is grounded or connected to the negative terminal. Hence special care must be taken to protect the circuit from accidental short circuits. Suitable enclosure must be used. Power should be taken from the battery directly before the ignition switch and it is better to use 7555.

Parts

Item	No. reqd	Description	Designation
1	1	.1uF	C1
2	1	250uF	C2
3	1	.01uF	C3
4	3	1K	R1, R3, R4
5	1	3.3K	R2
6	1	8	SPK1
7	1	555 Ohms	U1
8.	1	Piece of veroboard	
9.	1	D1	10V Zener diode
10.	1	D2	LED
11.	1	Q1	BC 147 Transistor

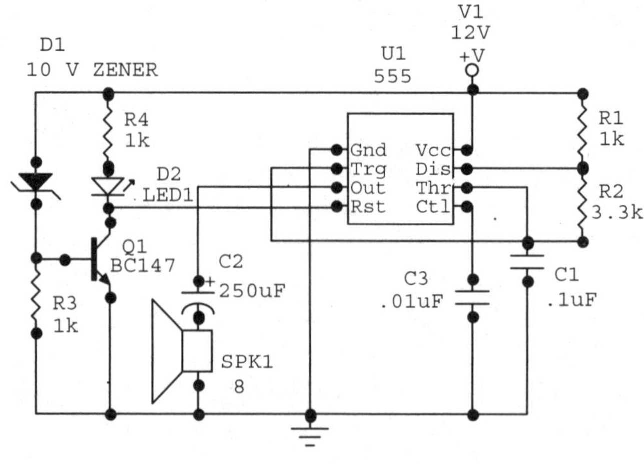

Schematic 14

Now that we have of enough of 555, let us have some projects with some other ICs

Fridge Door Alarm

Introduction

Normally we just push the fridge door close, hoping that the door gasket magnets will hold it. Often the door does not stay close tight in many households. Children do not generally bother to keep the door properly closed. If the fridge door is not kept closed properly, the cooling will be lost and fridge tends to work more, thus increasing your power bill. The circuit here reminds you with an alarm if the door is left open. Alarm continues to sound as long as the fridge door is open, telling you to do your job fast and close the door.

Description

We now make use of CMOS IC CD 4060. It is an interesting IC, which has a built in oscillator and divider. The basic oscillator can be configured either by RC or crystal network. Then CD4060 divides this oscillator frequency into binary divisions of 4, 5, 6, 7, 8, 9, 10, 12, 13, and 14 which are available as outputs. This means that IC can divide the basic frequency up to a maximum of 16384 times. Because of the division of basic frequency in large numbers, the timings are more stable. For normal operation reset pin is pulled low. If the reset pin is held high, all the counters are reset and stops oscillator. Oscillator Block diagram is shown in **Figure 20** and pin out of the IC is given in **Figure 21**.

Basic frequency is determined by the following formula.

$$F = 1/2.2\ R_1 \times C_X \text{ at } V_{CC}=10V$$

where R_2 must be greater than $2R_1$ to $10R_1$

OSCILLATOR BLOCK DIAGRAM

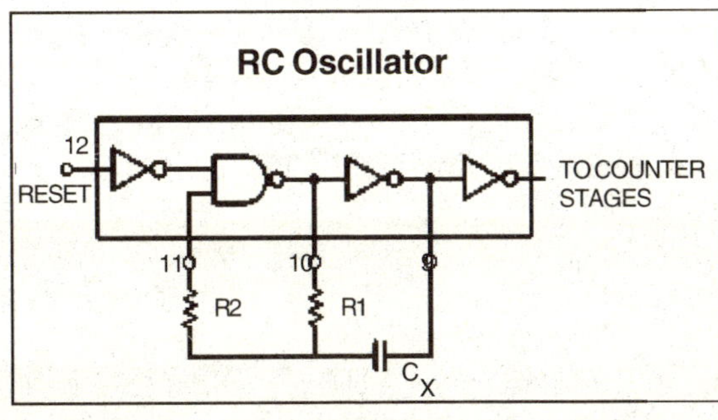

Figure 20

PIN OUT

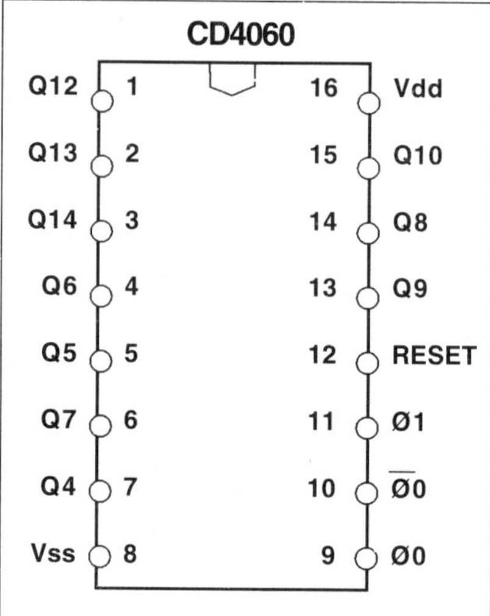

Figure 21

The circuit is shown in **Schematic 15.** When the fridge door is closed, the interior will be dark. Hence the LDR has a high resistance and the pin 12 (RESET) of the IC is held high. But if the door is left open, outside light enters the fridge or the fridge light itself comes up. This effectively lowers the resistance of LDR and Pin 12 goes low. Then IC 4060 starts oscillating and counting. With the present timing components at pins 9, 10, 11, the output timing is two seconds. Hence pin 3 goes high for two seconds and goes low for another two seconds. Three terminal piezo buzzer there is activated for every two seconds and alarm sounds until the door is closed. The circuit operates at 3V can even be used at 6V. You may use two pen cells in two cell plastic holder or four in a four cell holder. But if you wish to use NiCad cells you have to use at least three giving 3.6V. Cell voltage of NiCad cells is only 1.2.

Construction

Make this circuit on a small Vero board; fix it in a small box leaving a little window for the LDR and a small opening for the buzzer sound. LDR should point towards the light source. Time of oscillation can be adjusted by changing the values of C1, R1 and reduced by half by connecting the outputs to Pin 2. Switch is avoided to reduce the size. After all, the current consumption is low and batteries can be removed when not in use. Please do not place this in the freezer and make sure to remove this while defrosting. Connect the trigger terminal of the three terminal Piezo buzzer to Pin 3 of CD 4060.

Note

Please read notes at the end of the book. There are notes on care of ICs particularly CMOS ICs, tips and tricks and soldering techniques.

Note that CD 4060 is a CMOS IC. CMOS ICs can be easily damaged by the leakage currents and even by the static electricity picked by human body. Normal nichrome wire soldering iron has enough leakage current to affect this damage. In the good olden days, it was a practice to remove the hot soldering iron from the wall socket and solder CMOS IC pins. Presently most of CD series ICs are diode protected and can handle a bit of leakage transients. But prevention is always better than cure. But one must remember, "Cure of course is a spare IC!" Hence take all the precautions necessary for CMOS ICs.

Parts

Item	*No. reqd*	*Description*	*Designation*
1	1	Three terminal Buzzer	BZ1 BZ1
2	1	.01uF	C1
3	1	68K	R1
4	3	150K	R2
5	1	100K Preset	R3
6	1	LDR	R4
7	2	1.5 V battery cells	
	Or		
	3	Ni.Cad or Ni Mh cells	
8	1	Battery holder to suit above	
9	1	Piece of Veroboard	

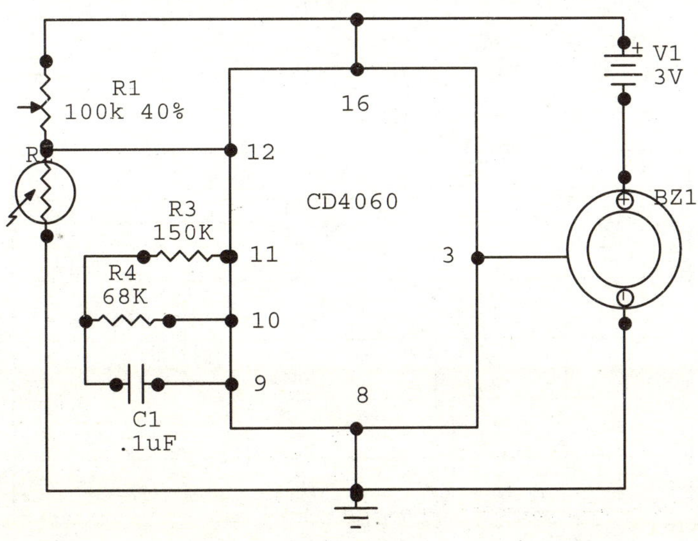

Schematic 15

Refrigerator Defrost Timer

Introduction

It is an extremely useful home project. Most of the people who use the fridge never bother about defrosting it. I have seen hills of ice in most of the freezers, more so in the humid areas. Ice on the freezer box makes the fridge work more and more. Being a bad conductor of heat, ice does not allow the cooling to be passed on further down, which makes the fridge inefficient. This circuit actually saves power by switching off the fridge at the desired timings and also thereby defrosting it.

The mains supply of the refrigerator is connected through this project. After every 20 hours, it disconnects the power for about two hours and reconnects again. The cycle goes on. This keeps the fridge always defrosted. Defrosting time can be changed, as also the actual operation.

Description

This circuit makes use of two ICs. *CD4541 is the first integrated circuit (U1), which has a built in oscillator and a programmable divider up to 65536 times. It provides the binary outputs of* 2^8, 2^{10}, 2^{13}, 2^{16}, *which are digitally controlled. It has an internal low power oscillator going up to 100 KHz but can also work as a single transition timer. It features automatic power on reset. Out put can be selected as normally high or low.*

RC oscillator frequency is determined by external RC network with the following formula.

$$F = 1 / (2.3 \times R_{TC} \times C_{TC}) \text{ when } R_S \text{ is} = 2R_{TC} \text{ where } R_S \geq 10K.$$

If R_{TC} *(R1) is 100k and* R_S*(R2) is 47k and* C_{TS}*(C1) is 0.47mF, time at the output is 2.09 hours, when* Q_{16} *output is selected. It will be 15 minutes if* Q_{13} *is selected. Division ratios can be selected from the following table. Present schematic shows the division ratio of* Q_{16}, *i.e., 65536 times. Division rates are more than CD4060 and IC has better control features. Division Ratio Table and Pin out details of CD4541 are given in* ***Figure 22*** *and* ***Figure 23*** *respectively.*

DIVISION RATIO TABLE			
A	**B**	**No. of Counter Stages**	**Count**
0	0	13	8192
0	1	10	1024
1	0	8	256
1	1	16	65536

Figure 22

PIN OUTS

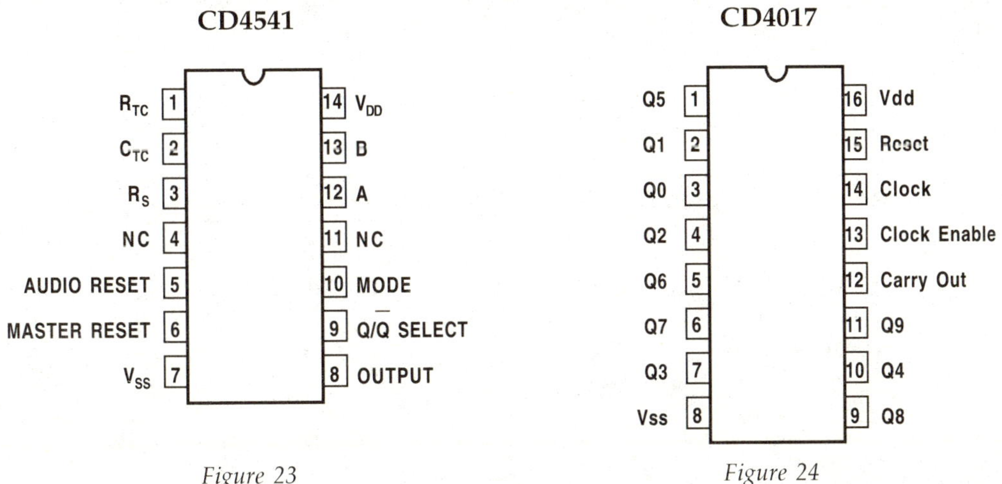

Figure 23 *Figure 24*

The output of this IC (two hour timing) is clocked into second IC, CD4017.

CD 4017 is a CMOS decade counter also known as Johnson ring counter. This IC divides the input clock pulses at Pin 1 by ten and all the ten divisions from Q0 to Q9 are available as outputs. Outputs are available as 0, 1, 2, 3, 4, 5, 6, 7, 8, and 9. For normal operation clock inhibit (13) and reset pin (15) are grounded. The counter advances one pulse at every positive going pulse at the input. For every count the corresponding decoded output goes high and all other outputs remain low. Carry out pin (12) is used for cascading more number of ICs for more number of divisions. It goes high for 5 pulses and goes low for another five pulses. Pin out details of CD4017 are given below.

With the first pulse from CD4541, Q_0 of CD4017 will be high for about two hours and goes low. This drives the relay RLY1 through D1, R3, and Q1 transistor. Fridge is connected through normally closed contacts of the relay, which means that the fridge is connected normally. The relay contacts should be capable of handling the fridge currents. As the relay fires, the contacts open and the power is removed from the refrigerator. It will stay thus for one clock period, i.e., 2 hours, after which relay falls back and the normally closed contacts close again feeding power back. Then Q_1 will be high for two hours and goes low afterwards and similarly all the ten outputs go high and low for 10 cycles until Q_0 becomes high again. Thus the relay fires for every 20 hours. The cycle continues. However there is an option for extending this cycle time by another clock period by closing the switch S2 which connects the Q_1 output through diode D2, to the relay drive transistor. Clock period reaching CD 4017 can be adjusted at CD4541. The circuit is shown in **Schematic 16.**

Construction

Both ICs are CMOS devices and should be used with respect as already suggested. General Vero boards will do well as it would be difficult to get a printed board designed for this.

House keeping is important. A good 12V power supply is recommended. Pin out of both ICs is given below. Reset pins of both ICs are connected to ground by R3. When pulled high by reset switch S1, the circuit resets.

Parts

Item	No. reqd	Description	Designation
1	1	.47uF	C1
2	3	1N4003	D1, D2, D3
3	1	BEL187	Q1
4	2	100K	R1, R2
5	3	10K	R3, R4, R5
6	1	Relay 12V, 300 OHMS	RLY1
7	1	Push to on switch	S1
8	1	SPST Switch	S2
9	1	CD4017	U1
10	1	CD4541	U2

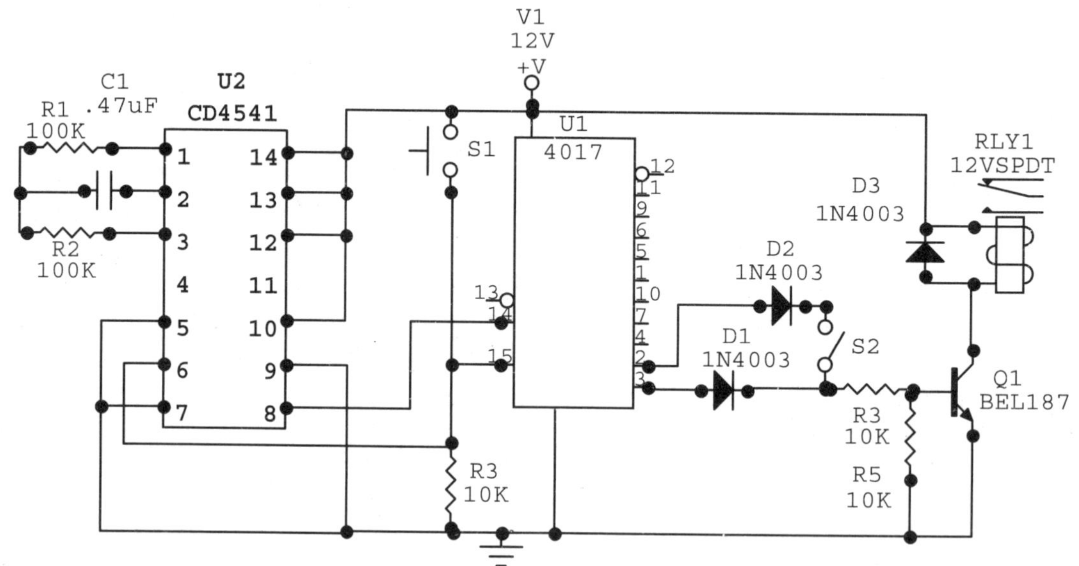

Schematic 16

Pseudo Random Lights

Introduction

Here is an interesting application for random light display. With this circuit as many as 72 LEDs of different colors can be used. LEDs can be fixed at random to suggest lighting them at random. They can be fixed one after the other suggesting running lights. They can be fixed in rows giving an effect of layers. It is an ideal choice for Christmas Tree Lighting and other festival lighting such as Divali. You can also fix them around statues, arches or banners. Imagination is unlimited.

Description

CD 4020 is also a 14-stage divider IC, as it divides clock pulses by 16384 times at the final output. Similar to CD 4060, but this IC has no built in oscillator; the clock pulses from an external source are to be fed into the input of this IC. Basic pulses are also available at Q_0 at pin 9. All but Q_1, Q_2 divisions are available as outputs. For instance Q_3 pulse division i.e., after dividing 8 times is available at pin 7. It is similar to CD4060, but CD4060 has built in oscillator whereas for CD4020, external clock should be supplied. Pin out detail of CD4020 is given in ***Figure 25.***

CD4020 is selected because an external clock can be given from the mains frequency as this circuit may always work off mains. Hence separate timing components are not required.

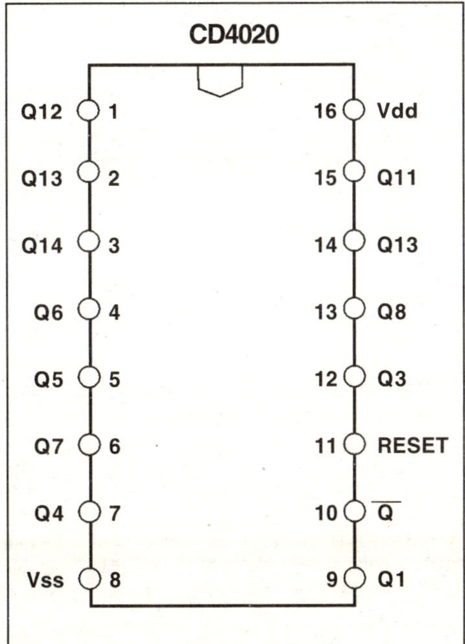

Figure 25

50Hz clock input is taken from mains supply. So 6.25 Hz is available at Pin 3, 3.125 Hz at Q_4 and so on. Each of these outputs drives a series of 6 LEDs at 10mA. LEDs light up in the order of Q_0, Q_3, Q4 ....Q_{14} and so on but not at random. 12 outputs are shown with the exception of Q1 and Q2. LEDs at the lower end flicker fast and those at the higher end stay on for long and go off for long.

Now the trick is to place these LEDs at different places to get a look of random distribution. Even LEDs within the same series are positioned at different places. Unlike other circuits, this circuit is shown along with a half wave rectifier and transformer, as reference frequency of 50 Hz is given through a 10k resistance (R1) from the secondary of transformer to CD 4020. Reset Pin 11 is pulled down by R2 10k resistance. The counter stops if it is pulled up. The circuit is shown in **Schematic 17.** Switch is not shown in the schematic.

Construction

Again this is a CMOS IC and you are warned. You may have to hard wire 72 LEDs around. Hence please take care not to make shorts. Outputs can handle only 10mA of current and so do not compromise on LED series resistance (R3 —R14) for more brightness. Do not use different colored LEDs in the same series, as they will not light up properly.

Note

Other ICs like CD4060, 4040, 4541 also can be used for this application. Make suitable changes following the appropriate pin-out and application notes.

Parts

Item	*No. reqd*	*Description*	*Designation*
1	1	1000uF	C1
2	72	LED	D1 to D72,
3	1	1N4003	D73
4	2	10K	R1, R2
5	12	1K	R3, to R14
6	1	230/12V Transformer	T1
7	1	CD 4020	U1

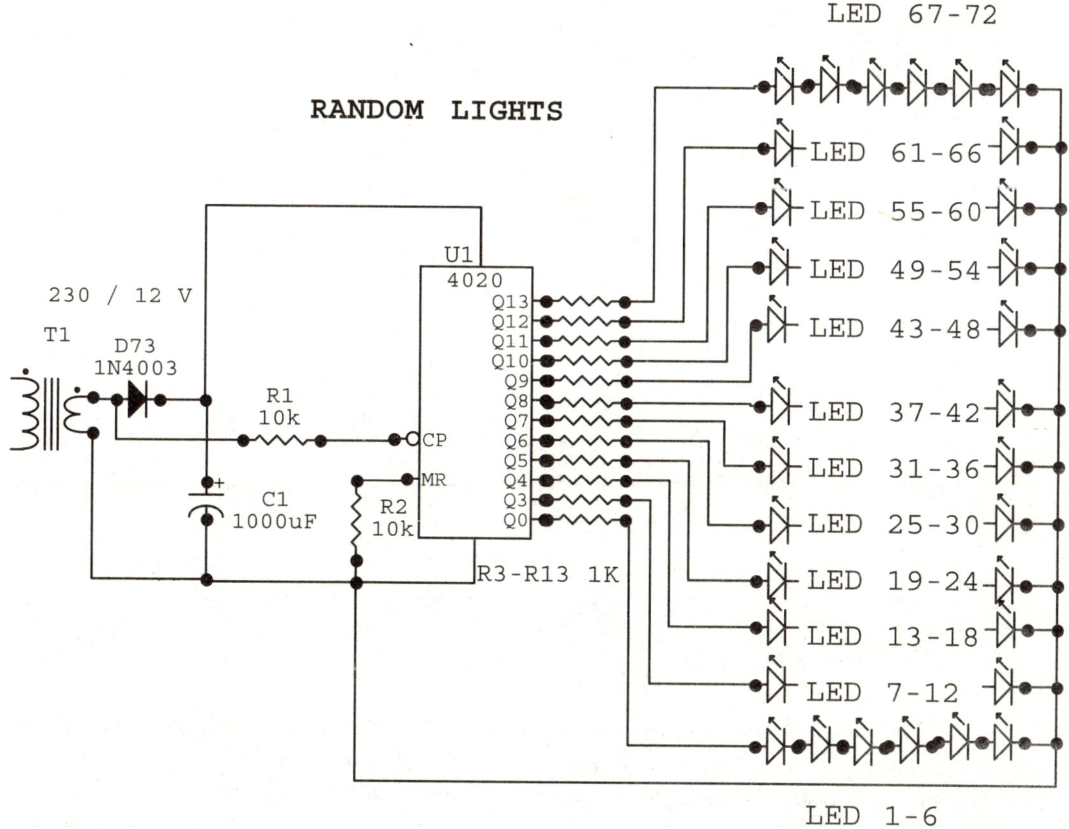

Schematic 17

Water Level Indicator

Introduction

We already had a good number of water projects and this is not just one more. This circuit gives continuous indication of six levels of water in the tank using ordinary wires as probes. Probe at the highest level can also give an alarm indicating water reached maximum level. The probe at the lowest level gives another alarm indicating low level. All the levels can be individually set and alarm set points also can be changed.

Description

This circuit is built around IC CD 4050 (Hex BUFFER). Pin out of the IC is given in ***Figure 26*** and the circuit in **Schematic 18.**

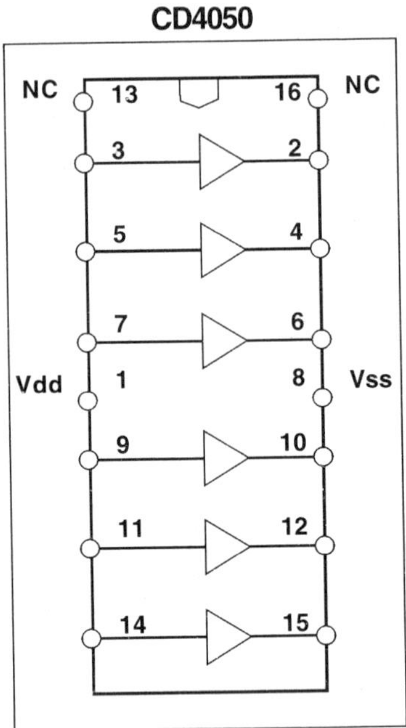

Figure 26

CD4050 has six separate buffers in a single package. In a buffer, if the input level goes high, output also goes high. Similarly if the input level goes low, output also goes low. On the contrary, in an inverter (CD4049) if the input level goes high, output goes low. Similarly if the input level goes low, output goes high.

The five input probes are just wires fixed at appropriate levels in the water tank and the wire from the POWER SUPPLY LINE is fixed at the bottom most position as a reference. All the inputs are kept low through 3.3M resistance (R1-R6) and the output pins will be all low as long as there is no water level at any probe. Hence level indicating LEDs do not glow.

As soon as water touches any probe, corresponding input reaches high level and its output also turns high. This results in glowing of respective level indicating LED in turn. Probe at U_{1A} pin is at the lowest level. When the water level falls below this, corresponding LED 1 will not glow.

Probe of U_{1B} pin is placed at one step above the lowest level so that it gives an advance indication of the impending low level. As long as the water is at this level, green LED 2 glows. If the water level falls below this, this LED no longer glows. As there is now a low level at this output, Q_1 goes into conduction and drives Piezo buzzer BZ 1. It continues to ring until water level touches this probe.

Probes of U_{1F} pins are at the highest level. Whenever water touches high level probe here, LED6 glows indicating high-level water. This also triggers the transistor Q_2, and buzzer BZ2 goes on. It is essential that the tones of both the buzzers are different such as continuous, intermittent or musical tones to give different indications for high and low levels.

Construction

This circuit can be operated on 12V power supply, even half wave. Use of batteries is also good. Probes can be simple hook up wires with the ends scratched. You may have to clean them when occasionally they get coated.

Note

It will be a good idea to place a probe at one step above the bottom most level. It gives an advance alarm indicating that the level is just about to drop down. Water tanks go empty just as the guests arrive or when we need water badly. To add to our travails, power also fails at that time as if design. This idea gives time to fill the tank early. Of course there is an LED indication at the bottom most level.

Other ICs like CD4069, 4049, also can be used for this application. Make suitable changes following the appropriate pin-out and application notes.

Parts

Item	*No. reqd*	*Description*	*Designation*
1	2	PEIZO BUZZER	BZ1, BZ2
2	1	RED LED	D1
3	1	ORANGE LED	D2
4	2	YELLOW LED	D3, D4
5	2	GREEN LED	D5, D6
6	1	BC157	Q1
7	1	BC147	Q2
8	6	3.3M	R1, R2, R3, R4, R5, R6
9	6	1K	R7, R10, R11, R12, R13, R14
10	2	10K	R8, R9
11	1	CD 4050	U1

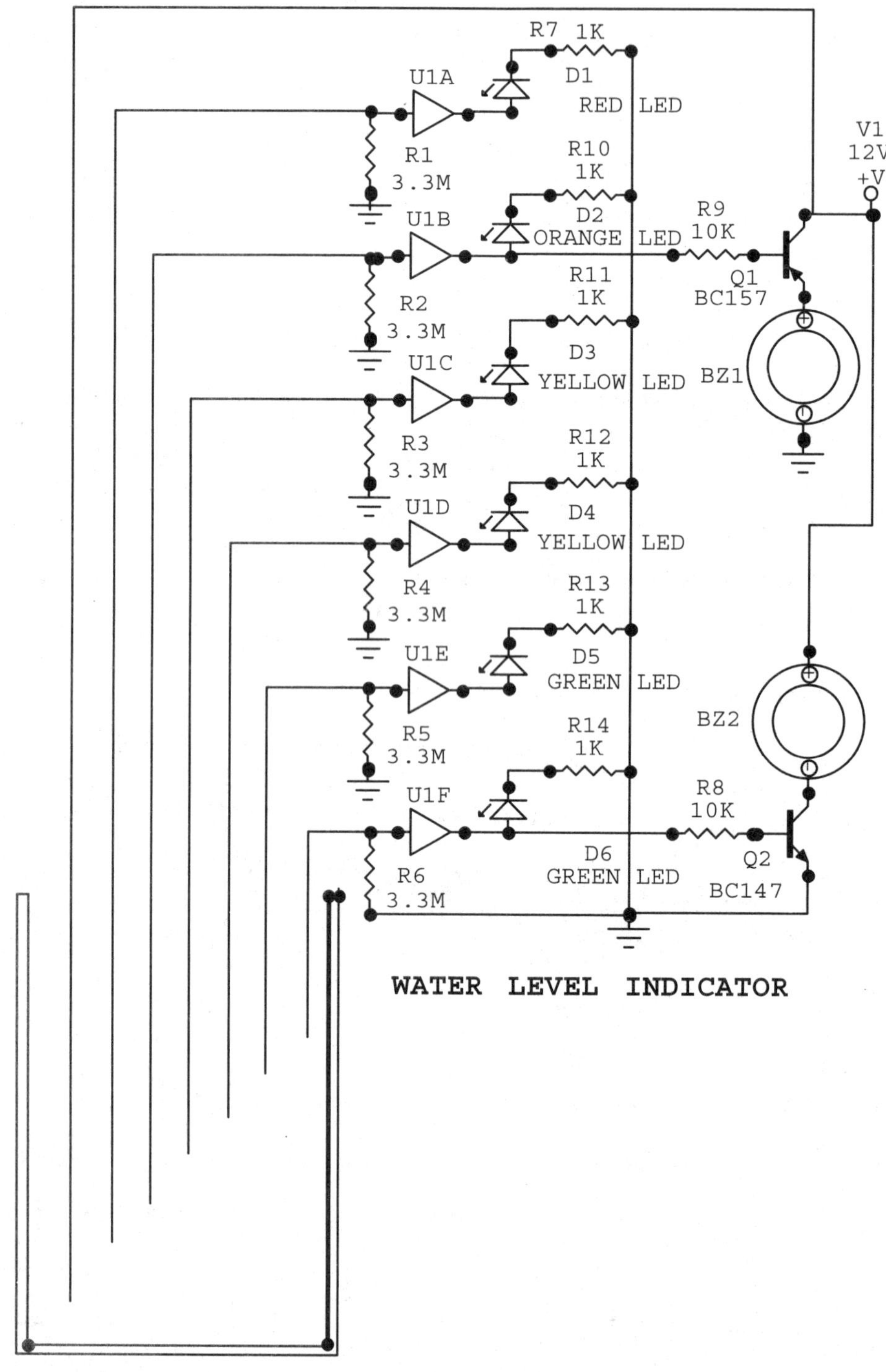

Schematic 18

Advertisement Display

Introduction

Here we have an unique application for an advertisement display such as "WELCOME" "HAPPY BIRTHDAY" or a shop display. It can be used a window display of a product or simply as a name board. It can be used as warning display at "ENTRY' or "EXIT" places or at manholes or "MEN AT WORK." Any letter in any language can be configured into a series of light bulbs operated at mains.

It does not use microprocessors and soft wares or EPROMs but a simple TTL shift register chip known as 74 LS 164. For the display it uses conventional series set operated at mains supply but LEDs also can be used. (LED scheme is not given here) Triacs are used for direct AC switching and hence relays are dispensed with. Each letter glows one after the other, stays on until all the letters are on. All of them go off at the last clock pulse and stay off for one clock pulse.

If you do not wish to use letters as display, you may use the series sets as they also can be configured around various parts of a statue or a display item. It gives a dynamic effect as the lights come up one after the other. Speed of the display movement can be easily adjusted. Creativity is limitless!! You may use it to display your own name plate using LEDs. Wonderful it will be as the moving letters attract attention!

Description

74LS164 is a serial in parallel out shift register. In this IC, data coming serially into the input of the IC are split, registered into 8 bits and available as parallel outputs. Hence the serial data is effectively split into parallel bits. ICs are also available with designations of Parallel In- Parallel Out, Parallel In -Serial Out. Pin out of 74164 is given in ***Figure 27*** *and triac configuration in* ***Figure 28.***

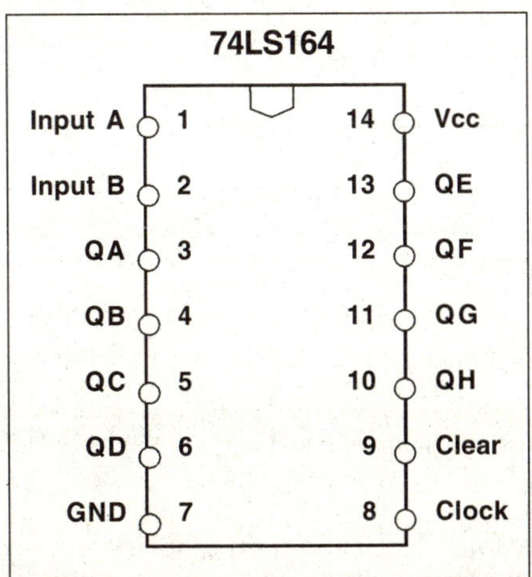

Figure 27

The circuit is shown in **Schematic 19.** IC1 555 generates the clock pulses and are fed serially, i.e., one after the other into it at Pin 8 of IC2. The frequency of these pulses is adjustable by R18 preset potentiometer. IC2 registers these pulses in parallel and releases to outputs at Q_0, Q_1......Q_7 up to 8 pulses. In this case Q_0, Q_1, ...Q_7 goes high and stay that way until reset. These pulses drive transistors Q_1 to Q_7, which in turn fire triacs Q_9 to Q_{15}. As these pulses are now connected to mains 230V AC, by the action of triacs, light bulbs glow one after another.

You can drive a series of lights as different letter of a message, for instance WELCOME. Q_9 Triac drives a serial set configured as W, Q_{10} drives E and so on. After all the letters are lit up the display goes blank for one pulse time and starts all over again.

Speed of these pulses can be adjusted at IC 1 with R18. Q_8 actually resets the IC at every 8th pulse so that the display starts all over again. If you want only 4 letters are to be displayed, shift the reset to 5th output. You can cascade more ICs to get more and more letters on display. But the limit of output current is 10 mA for any segment and this has to be strictly maintained.

Notes on Triacs

The triac is a three terminal semiconductor for controlling current in either direction. If you notice the symbol, you will find that it looks like two SCRs in parallel (opposite direction) with one trigger or gate terminal. The power terminals are designated as Main Terminal1 (MT1) and Main Terminal2 (MT2). It can be treated as TRIGGER AC. Unlike SCR it can trigger in both directions with a small voltage applied at the gate. Minimum holding current, must be maintained in order to keep a triac conducting. Main terminals are not interchangeable.

TRIAC-CONFIGURATION

PINNING - TO220AB

PIN	DESCRIPTION
1	main terminal 1
2	main terminal 2
3	gate
tab	main terminal 2

PIN CONFIGURATION

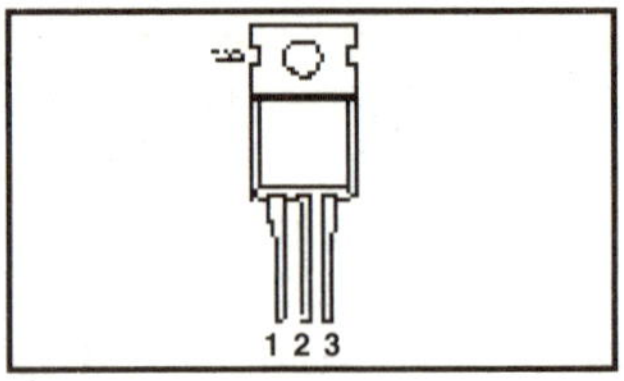

SYMBOL

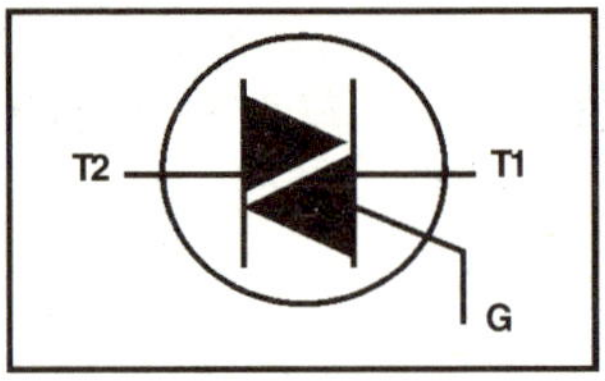

Figure 28

Construction

Entire circuit works on mains supply and you are forewarned. Now U1 IC is TTL version. Unlike CMOS ICs, static electricity does not pose a problem for TTL ICs but these cannot tolerate voltage variations. A regulated voltage of 5V is strictly required. Use 9V full wave rectifier and 7805 regulator.

In general ICs do not like bad housekeeping, more so when handling mains voltages. Soldering is straightforward. A piece of Vero board is OK. But if you are mounting triacs on the same board it is very important to have enough space between tracks such that high voltage arcs do not jump across. It is good idea to remove alternate tracks and mount triacs. 100K preset R18 is the speed control. BT 136 triacs can easily drive 500 W of power. Suitable heat sink must be firmly fixed for each triac individually. Mounting is similar to LM317.

Do not touch any part of the circuit while it is in operation.

Take for instance display of WELCOME. Take a hylum or plywood board of say, of 0.75 meter wide and of suitable length. Take a mains operated series set of bulbs and form the letter 'W' in the board and terminate. Then take another series set form it as letter 'E' and terminate. Similarly make the series sets for all other letters. You may drill and fix the bulbs in the forms of letters also. These terminals are connected to the outputs of the triacs Q9 to Q15. The other end of letter bulbs are made common and connected to the neutral side of mains.

Note

If you wish to use LEDs instead of direct mains bulbs, you may do so. It will not be a good idea to use LEDs at mains voltage at this stage. You may use 12 or 18 power supply, thus making the whole circuit safer to handle. (Keep the 5V regulator for ICs, nevertheless.) Remove the triacs. Connect LED in series marking the polarity. Number in a series depends on the voltage you wish to use and LED voltage. *For instance a red LED requires 1.7 Volts and green, orange and yellow needs about 2.0 Volts. And that is precisely the reason why you can not mix colors in the same series.* However you can connect a number of series of the same color in parallel and connect them to the output transistor keeping in mind BC 147 can handle 250 milliwatts of power. Use a current limiting resistor. BEL187 is a better choice which can handle about 1 W with a small heat sink.

Parts

Item	*No. reqd*	*Description*	*Designation*
1	2	.1uF	C1, C2
2	1	4.7uF	C3
3	7	100W	L1, L2, L3, L4, L5, L6, L7 (Light bulbs/series sets)
4	8	BC147	Q1, Q2, Q3, Q4, Q5, Q6, Q7, Q8
5	7	BT136	Q9, Q10, Q11, Q12, Q13, Q14, Q15,
6	9	2.2K	R1, R2, R3, R4, R5, R6, R7, R8, R9
7	7	330	R10, R11, R12, R13, R14, R15, R16
8	1	10K	R17
9	1	100KPreset	R18
10	1	74LS164	U1
11	1	555	U2

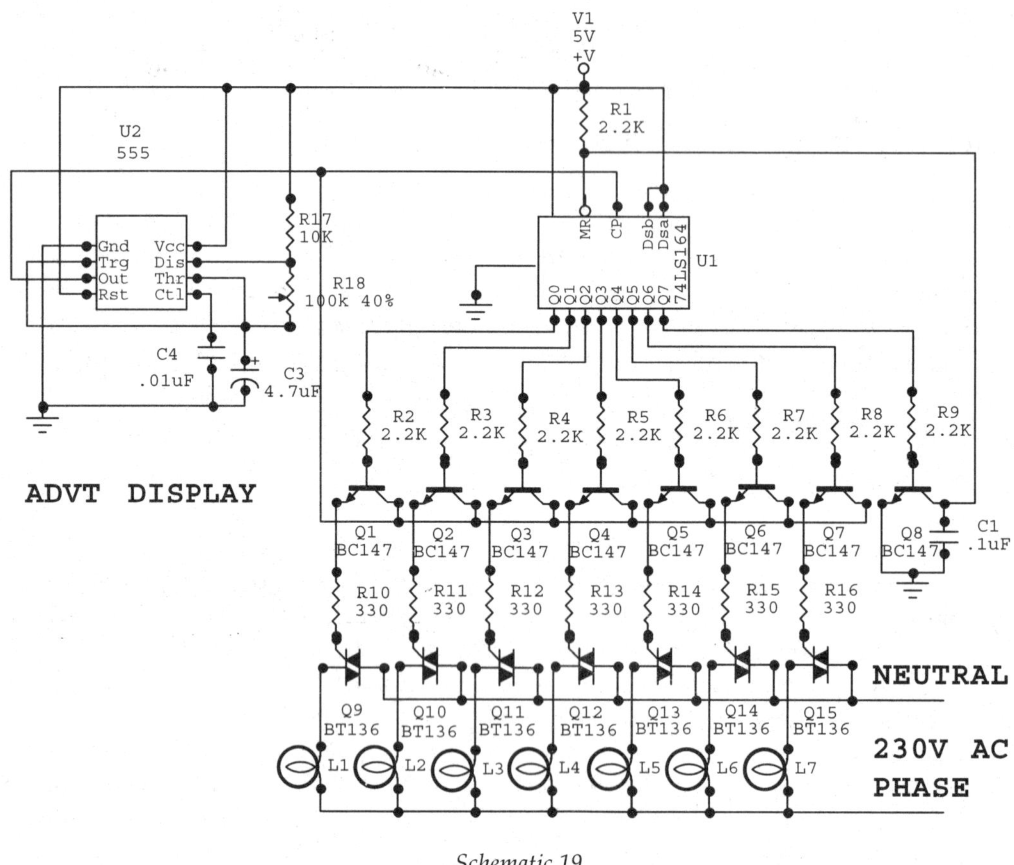

Schematic 19

30-minute Kitchen Timer

Introduction

Here is a kitchen need built around CD 4060. Often it so happens that the housewife forgets some thing left on the hot stove or in the washing machine as she watches TV or she talks to a friend over a neighborhood fence or she just wants to try a new recipe requiring a finite time. This is a timely personal reminder which gives an alarm, say after 30 minutes, or any other preset time. It can also work as a clock for the students who appear for the competitive exams.

Description

The circuit is shown in **Schematic 20.** Again CD 4060 is used, a 14 stage divider oscillator IC. Oscillator frequency is set by the components at Pins 9, 10, 11. Frequency can also be adjusted using variable resistance R1. With the components shown an alarm can be adjusted for 30 minutes at the output. Outputs at Q14 at Pin 3 and out put Q13 at Pin 2 are connected to the two terminals of SPST switch S2. Q14 gives 16384 divisions and Q13 gives 8192 divisions.

Switch S2 selects any one of the outputs and connects to BC147 transistor. When the selected output goes high, the transistor drives two terminal piezo buzzer. It also clamps CD 4060 from further oscillation through D3. If Q14 is selected timing will be 30 minutes and when Q13 is selected the timing will be 15 minutes as adjusted by R1. By using a multi pole single throw switch, other timings can be achieved. We can get division ratios of Q10 at pin 15, Q9 at pin 13 and Q8 at pin 14, which can be used appropriately. Use musical buzzer to give a pleasant sound in the kitchen.

Flashing LED at Q3 is incorporated to indicate the activity of the IC. S1 is the reset switch. When S1 is pressed even momentarily, Pin 12 goes high, it resets all counters and oscillator starts allover again.

Construction

Construction is straightforward and this IC is a CMOS device. Construction on a Vero board should do well. A simple full wave supply of 12V would suffice. During trials keep the output buzzer at Q10, or Q11. It gives lesser divisions and correspondingly less time cycle of operation. It is tedious to wait for half an hour to check the operation of the circuit during trials. This makes for a faster adjustment cycle time.

Other timings can be achieved by manipulating components at 9, 10, and 11 with the formula given earlier. As the basic frequency is divided a number of times, accuracy is better even if the time delay is in the order of few hours. 555 cannot achieve this accuracy easily.

Parts

Item	*No. reqd*	*Description*	*Designation*
1	1	Piezo Buzzer	BZ1
2	1	.1uF	C1
3	1	LED	D1
4	1	LED	D2
5	1	1N4148	D3,
6	1	BC147	Q1
7	1	1M variable	R1
8	2	47K	R2, R4
9	1	1M	R3
10	1	1K	R5, R6
11	1	10K	R7
12	1	Push to on switch	S1
13	1	SPST switch	S2
14	1	CD4060	U1

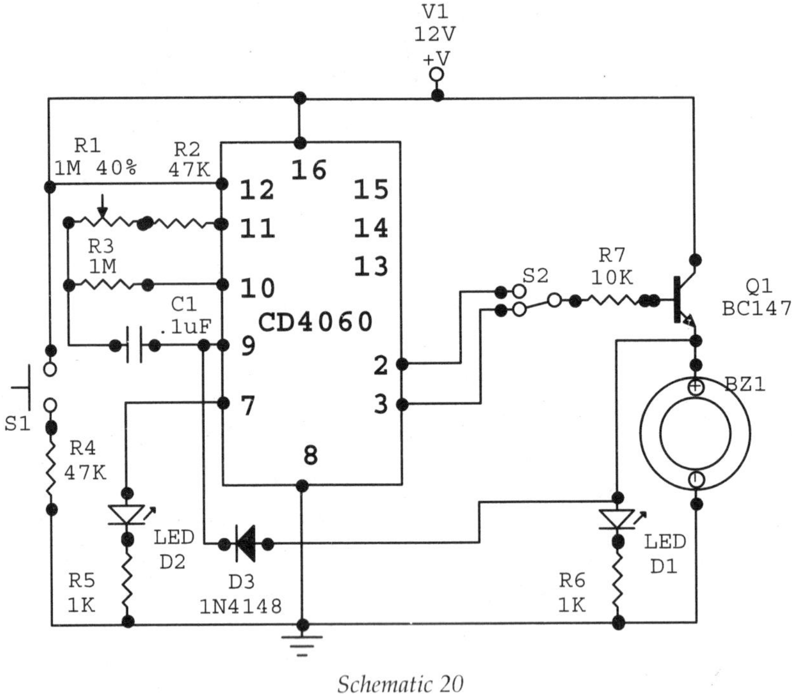

Schematic 20

Touch Switch With CD4066

Introduction

How interesting it will be if the lights come on when we touch a spot and goes off when we touch that again. This also does away with the mechanical contacts, switches and we can control different appliances from a single place. Here we have circuit for four individual touch switches, which can be used for different applications such as lights, fans, etc. With one touch, the relay fires, contacts close and another touch opens the contacts. They are fashionable, concealed and replace bulky switches. It adds a novelty and beauty in the interior decoration. As many as four switches can be configured from a single IC. Touch sensors could be just small dots of conducting plates, say copper, just three dots for each sensor. Relay contacts can be used in place of fan regulator contacts to make a touch control of fan speed.

Description

Using CMOS device for the job simplifies the design, as they can be triggered at extremely low currents. Their normal current consumption is also very low making them ideal for continuous operation. *CD 4066 is such an interesting CMOS switch IC, which has four individual bilateral switches inside the package, each capable of switching analogue or digital signals.* Pin assignment of CD 4066 is given in ***Figure 29.***

PIN OUT

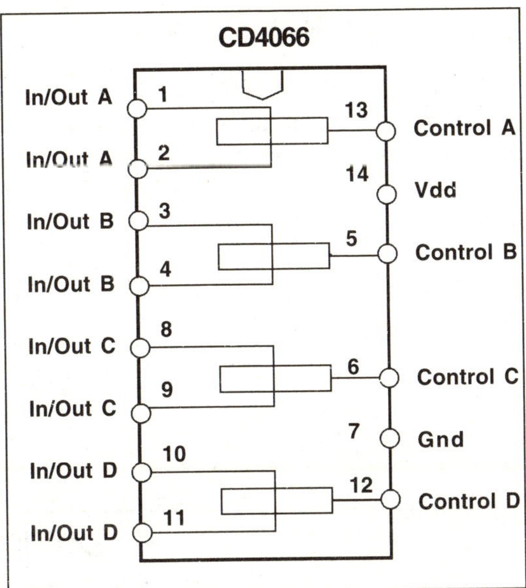

Figure 29

Take for instance one single switch (A) at Pins 1, 2 and 13. Pins 1 and 2 are switch contacts and Pin13 is the control terminal. As long as the control terminal (13) is low, there will be infinite resistance between 1 and 2 pins, which means that the circuit is open. If the 13 Pin goes high, the contacts close and there will be negligible resistance across them. If it goes low, the contacts open again.

Switches can operate at high speed up to a minimum frequency of 40 MHz. However they can only handle 10mA of current, which must necessarily be amplified for higher currents. All the other three switches at (B) Pins 3, 4 and 5, (C) 6, 9 and 8, and (D) 12, 10 and 11 are similar and independent. Being CMOS devices they need extremely small currents to operate the switches (10nA is typical). This feature is made use of in making the touch switches. We can make use of a single IC to control four different switches. All the four switches act similarly and independently.

Touch plates are three contact plates, which can be very small dots of 1mm copper but adequately spaced. One dot is connected to positive supply rail and center one is connected to the control terminal through a 100K resistance (R3, R7, R11, and R15). Third one is connected to the ground.

Now when fingers touch control dot and positive dot, the control terminal goes high and brings the required switch into conduction. Control switch is kept in high state by a feedback of 10M resistor (R4, R8, R12, and R16). Switch now drives a transistor (Q_1, Q_2, Q_3, and Q4), which fires a corresponding relay. LEDs (D1, D3, D5, and D7) show which of the switch has been closed. But now when we touch the control dot and ground dot, control pin is pulled down and switch goes off. Control pin stays low with the feedback.

Construction

CMOS DEVICE!!!!!!

If any of the inputs is not used, it is essential to connect that unused input terminal to ground to save the IC from getting damaged. Please note that this is a standard practice with CMOS ICs. Do not ground outputs. Construction on a Vero board is adequate but keep the tracks and gaps neat and clean in view of high impedance device. Wet flux can create havoc. In view of the shock hazard, relays are taken as the output devices. 12V full wave power supply with a good transformer is adequate. Mains terminal in and out of the circuit can be taken via a suitable strip block and connected to various appliances. Relay contacts should be rated at the required current. There should be adequate distance between each other touch contact to prevent wrong touches and triggering.

Parts

Item	*No. reqd*	*Description*	*Designation*
1	4	1N4003	D2, D4, D6, D8
2	1	LED1, 2, 3, 4	D1, D3, D5, D7
3	4	BC187	Q1, Q2, Q3, Q4
4	4	10K	R1, R5, R9, R13
5	4	1K	R2, R6, R10, R14
6	4	100K	R3, R7, R11, R15
7	4	10M	R4, R8, R12, R16
8	4	Relay 12V 300 OHMS	RLY1, RLY2, RLY3, RLY4
9	1	CD4066	U1

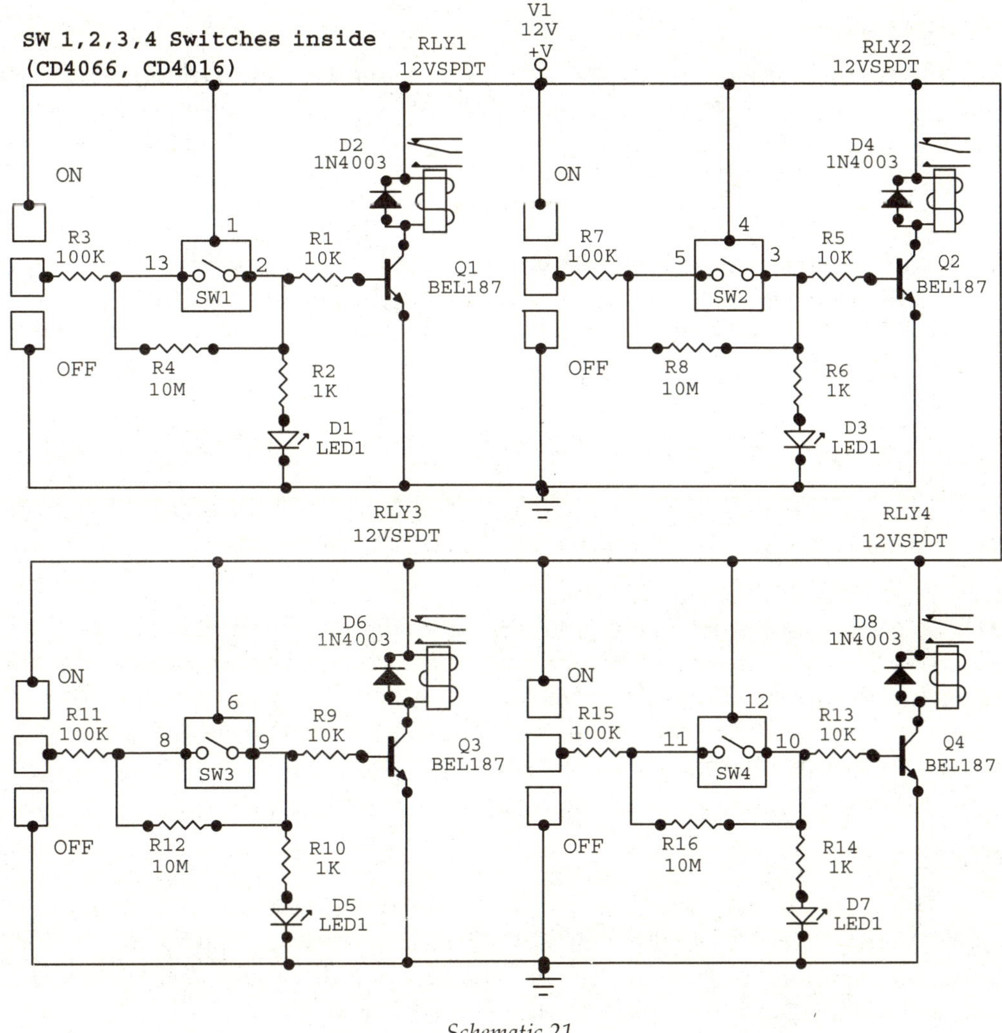

Schematic 21

Turn Signal Indicator

Introduction

Generally cars have right turn and left turn signals which are given out by blinking 12V bulbs fixed at the rear of car. These are operated by hand control at the steering wheel. To make them effective, quite a lot of power is used up. Still it so happens that the drivers at the rear may not make note of this, as they are a bit way down the car. We have here a novel, cost effective and energy efficient signaling display, which can fixed near the rear window or so that it becomes extremely visible.

Description

The circuit is shown in **Schematic 22.** This is another application of CD4017, decade counter. Two identical circuits are built here one for each direction. The advantage is that they can be built straight into the left; right signal lights and power can be taken from them without additional wiring from the steering wheel.

Clock pulses to the IC are supplied by good old 555, which can be adjusted by 100K (R2) preset. CD 4017 divides these pulses and outputs are available at Q_0 to Q_9. Each output drives a Red LED. All these ten LEDs are fashioned as an arrow. The other section of the circuit is similar with R5 is the speed control. The LEDs on one circuit show an arrow in the direction of right and vice versa.

Construction

The circuits can be built on a single Vero board or two pieces of Vero boards for each of right and left signals. Power is taken from the lights itself appropriately.

Arrows with LEDs can be fixed on transparent acrylic sheet of suitable size. Drilling holes for LEDs should not be a problem but without a little care, acrylic sheet may break. By using an acrylic sheet, view of the rear is not lost even if the display is placed there. You may use more LED (up to 6 for red LEDs) in series for each output, but the current limiting resistor (R4, R8) must be properly trimmed. Output current should not exceed 10mA. R2, R5 presets are speed controls for LEDs.

Note

Any one of these circuits can very well be used for window display lights, running lights and even as an electronic dice. When these LEDs are fixed concentrically in two circles, the circuits can work as hobby roulette wheel, which could be very popular in school functions. However for these circuits, the clock frequency should be suitably increased.

Parts

Item	*No. reqd*	*Description*	*Designation*
1	2	47uF	C1, C4
2	2	.1uF	C2, C3
3	20	LED1 to 20	D1 to D20
4	4	10K	R1, R3, R6, R7
5	2	100K	R2, R5
6	2	1K	R4, R8
7	2	CD4017	U1, U3
8	2	555	U2, U4

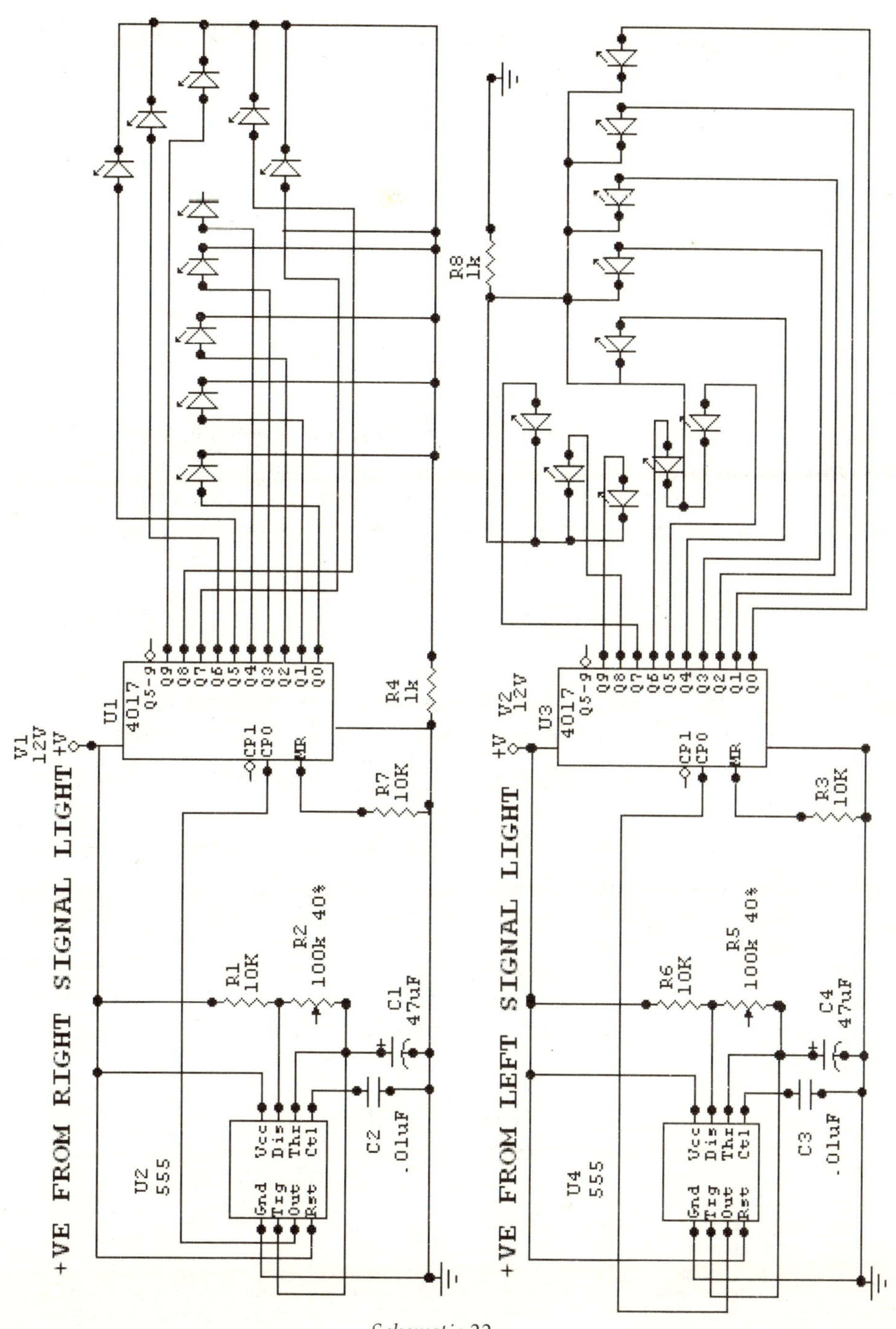

Schematic 22

Staircase Switch with CD4013

Introduction

If you really want to have classical staircase switch with push to switch on from one place and off from another place, here you have more circuits. Now you have two options to build the circuit here, one with light sensitive and another without it. Time delay is not incorporated here. We can use both facilities in a single IC or build them in separate ICs. Any or all of them can be used depending on the need.

Instead of ever present 555, this circuit now employs CD 4013 which has two D type latches in a single package. The circuit toggles a load with a momentary push button. Several push buttons can be fixed in parallel at various places to control the triac from any of those many locations. Two independent circuits can be made from a single IC. There are three options are here to show the flexibility of the idea and as a kind of tutorial for possible applications. Ideas from these circuits can be implemented in other similar circuits with due care.

Notes

A flip-flop is used to store or 'lock' one bit of information. This is known as 'latching.' Digital electronics and especially computers use a number of flip-flops, which latch several bits of data at the precise moment. There are a few variations of a classical flip flop i.e., JK, and RS. D type flip-flop is also one of them.

D-type flip-flop is just a clocked flip-flop with a single digital input D (D for Data). Every time a D-type flip-flop is clocked, its output follows whatever the state of D is in. Intermediate state is avoided in this flip-flop. When the clock goes high, data at D (0 or 1) is transferred to Q. $\overline{Q}$ *will have the opposite of this state. When clock goes low, data remains unchanged. Q stores data until the clock goes high again when new data may be available.*

CD4013 has two independent D type latches with set, reset, data and clock inputs. Both Q and are available as the outputs, which mean that both output states (high and low) or both toggle states are available. Set or reset is independent of clock. Truth table and Pin out of IC are given ***Figure 30 and Figure 31*** *respectively.*

TRUTH TABLE

CL†	D	R	S	Q	$\overline{Q}$
↑	0	0	0	0	1
↑	1	0	0	1	0
↓	x	0	0	Q	$\overline{Q}$
x	x	1	0	0	1
x	x	0	1	1	0
x	x	1	1	1	1

No change
† = Level change
x = Don't care case

Figure 30

PINOUT

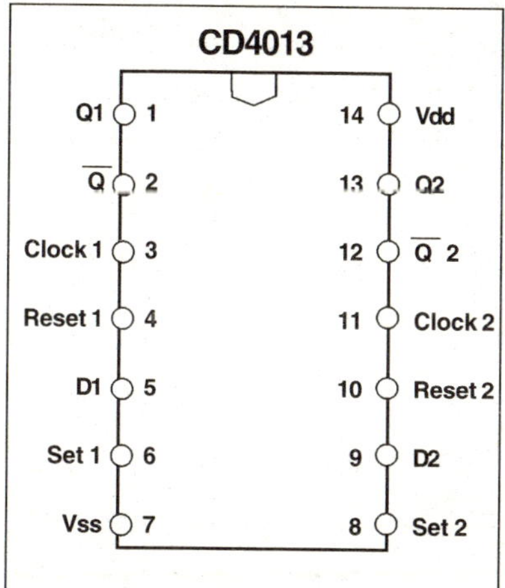

Figure 31

Description

Now we have a series of circuits here using this IC. Circuit 1 and 2 are the same but the second circuit makes use of triacs. Third circuit has the option of light sensitivity incorporated in to it. Use of triac permits operation directly at mains. This IC has two D type latches and the description holds good for both. Hence it is possible to make two independent switches from the same IC.

Circuit 1

The circuit is shown in **Schematic 23.** As soon as the switch S1 is pressed, clock input goes high and high data input is transferred to the output which drives a transistor, and hence forth a relay. Now $\overline{Q}$ out put will be low and is coupled to data input now. At the next switch on of S1 or S2, clock input goes high and data low at D input is transferred to Q_1 transistor. Now Q_1 cannot any longer hold the relay up, and equipment switches off. Now $\overline{Q}$ Output is at high level, which is coupled to data input and the latch is ready for next sequence.

Construction

In general ICs do not like bad housekeeping, more so when handling mains voltages. Soldering is straightforward. A piece of Vero board is OK. CD4013 is a CMOS IC. Please respect it. Relays should be rated at the current required. *D1 and D3 diodes are provided to protect the components from the back EMF generated by the relay coil. This is a standard method of relay circuit protection.* Only two switches are shown as examples. You may add more number of switches for use at a number of different locations.

Parts for Relay Circuit

Item	No. reqd	Description	Designation
1	2	1N4003	D1, D2
2	2	BEL187	Q1, Q2
3	2	10K	R1, R3
4	2	33K	R2, R4
5	2	150K	R5, R6
6	2	Relay 12V/300 OHMS	RLY1, RLY2
7	4	Push to On switches	S1, S2, S3, S4
8	2	CD4013	U1A, U1B

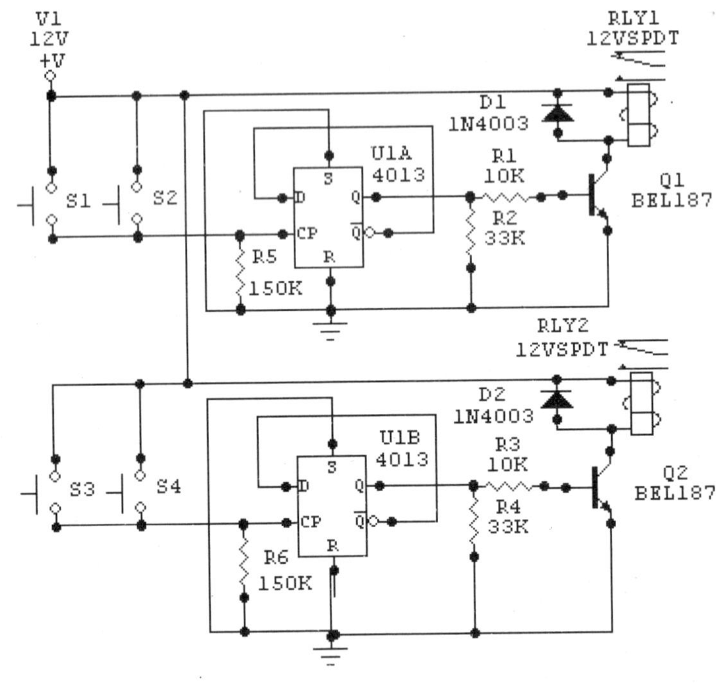

Schematic 23

Circuit 2

Circuit 2 makes use of triacs and operation is similar, which is shown in **Schematic 24**. AC voltage is lowered by 10K /10W resistor (R8) and rectified by D2. The voltage is filtered by C1 and stabilized by zener diode D4. Entire earlier description holds good except now the output transistor Q1 drives a triac Q3 instead of a relay. The gate of triac is connected at the emitter of Q1 BEL 187 transistor. Similarly Q2 transistor drives Q4 triac as its gate connected at the emitter of Q2. There is an indication of LEDs also connected at these emitters. You may use different color LEDs at these places. When the triac fires, the light bulb glows. There is a detailed description about triacs in an earlier circuit.

Construction

In general ICs do not like bad housekeeping, more so when handling mains voltages. Soldering is straightforward. A piece of Vero board is OK. But if you are mounting triacs on the same board, or separately, it is very important to have enough space between tracks such that high voltage arcs do not jump across. It is good idea to remove alternate tracks and mount triacs. BT 136 triacs can easily drive 500 W of power. Suitable heat sink must be firmly fixed for each triac individually. CD4013 is a CMOS IC. Please respect it.

Only two switches are shown as examples. You may add more number of switches for use at a number of different locations.

You are warned as the entire circuit works off 230 V AC mains unlike the earlier circuit. If you wish to have isolation from mains use the relay circuit. Use of triac simplifies the circuit as it does away with the transformer power supply and removes the problem of relay holdup and chattering.

Parts List (Traic Circuit)

Item	*No. reqd*	*Description*	*Designation*
1	1	470uF	C1
2	2	LED	D1, D2
3	1	1N4007	D3
4	1	12V ZENER	D4
5	2	100W Bulbs	L1, L2
6	2	BEL187	Q1, Q2
7	2	BT136	Q3, Q4
8	2	10K	R1, R3
9	2	33K	R2, R4
10	2	150K	R5, R6
11	2	1K	R7, R9
12	1	10K/10W	R8
13	4	PUSH TO ON SWITCHES	S1, S2, S3, S4
14	2	CD4013	U1A, U1B

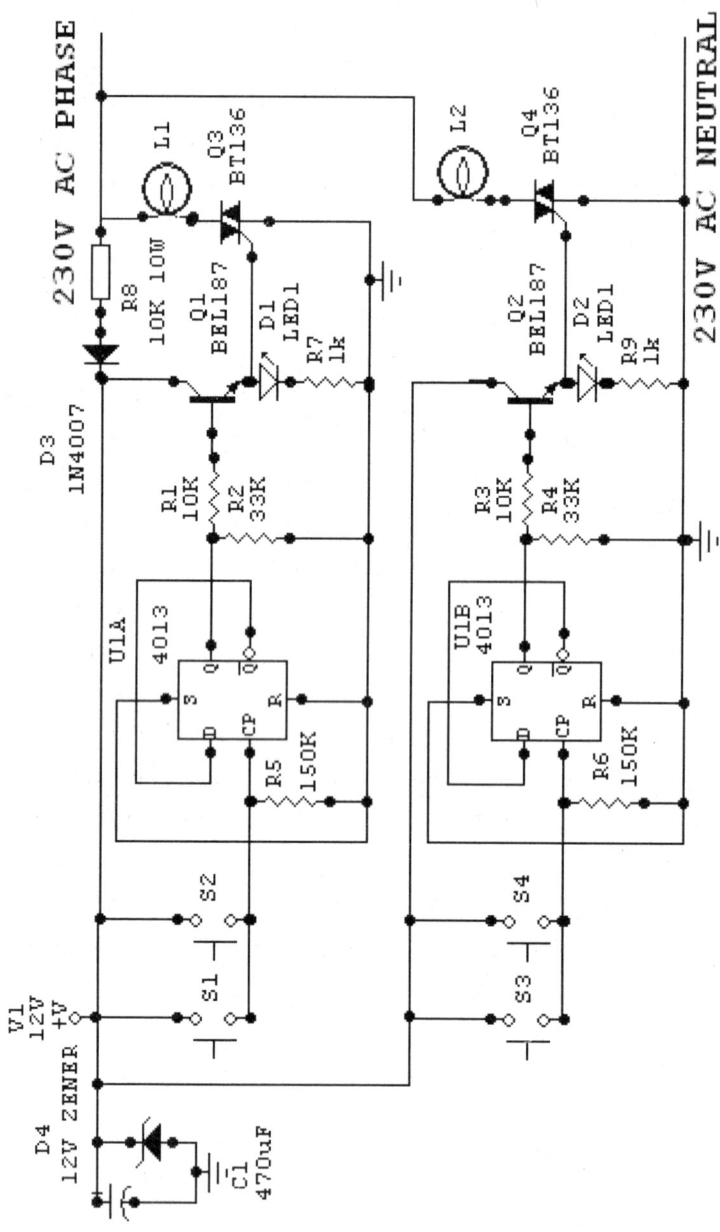

Schematic 24

Light Sensitive Staircase Switch with Triac

Circuit 3

Operation of the third circuit is much similar except that it has photo sensitivity. The circuit is shown in **Schematic 25.**

When there is no sufficient light, the reset pin is held low by R14 (4.7K resistor) and the circuit works as it is described in the earlier paragraph. But if there is sufficient light, resistance of LDR goes low and the transistor goes into conduction. Now the reset pin goes high and the circuit cannot work. R16 is the sensitivity adjustment. You may use a different color of LEDs at the emitter of Q1 and Q4. You may construct the circuit on a single board and can use it for two different applications with a single IC of CD4013. You can use same application in two different latches or make both the same.

Construction

In general ICs do not like bad housekeeping, more so when handling mains voltages. Soldering is straightforward. A piece of Vero board is OK. But if you are mounting triacs on the same board, or separately, it is very important to have enough space between tracks such that high voltage arcs do not jump across. It is good idea to remove alternate tracks and mount triacs. BT 136 triacs can easily drive 500 W of power. Suitable heat sink must be firmly fixed for each triac individually. Please respect CD4013, a CMOS IC.

Only two switches are shown as examples. You may add more number of switches for use at a number of different locations.

Please be careful that the entire circuit works off 230 V AC mains. If you wish to have isolation from the mains, use relays instead of triacs and use transformer power supply of 12V full wave. Light sense circuit consisting of Q5 and Q6 can be cleverly added to the relay circuit No.1

Parts

Item	*No. reqd*	*Description*	*Designation*
1	1	470uF	C1
2	2	LED1	D1, D5
3	1	1N4007	D2
4	1	12V ZENER	D4
5	2	100W Light Bulbs	L1, L2
6	2	BEL187	Q1, Q4
7	2	BT136	Q2, Q3
8	2	BC157	Q5, Q6
9	4	10K	R1, R3,

Contd...

Item	No. reqd	Description	Designation
10	2	33K	R2, R4
11	2	150K	R5, R6
12	2	1K	R7, R9
13	1	10K/10W	R8
14	2	4.7K	R14, R17
15	2	LDR	R12, R15
16	2	100k variable	R13, R16
17	4	Push to On switches	S1, S2, S3, S4
18	2	CD4013	U1A, U1B

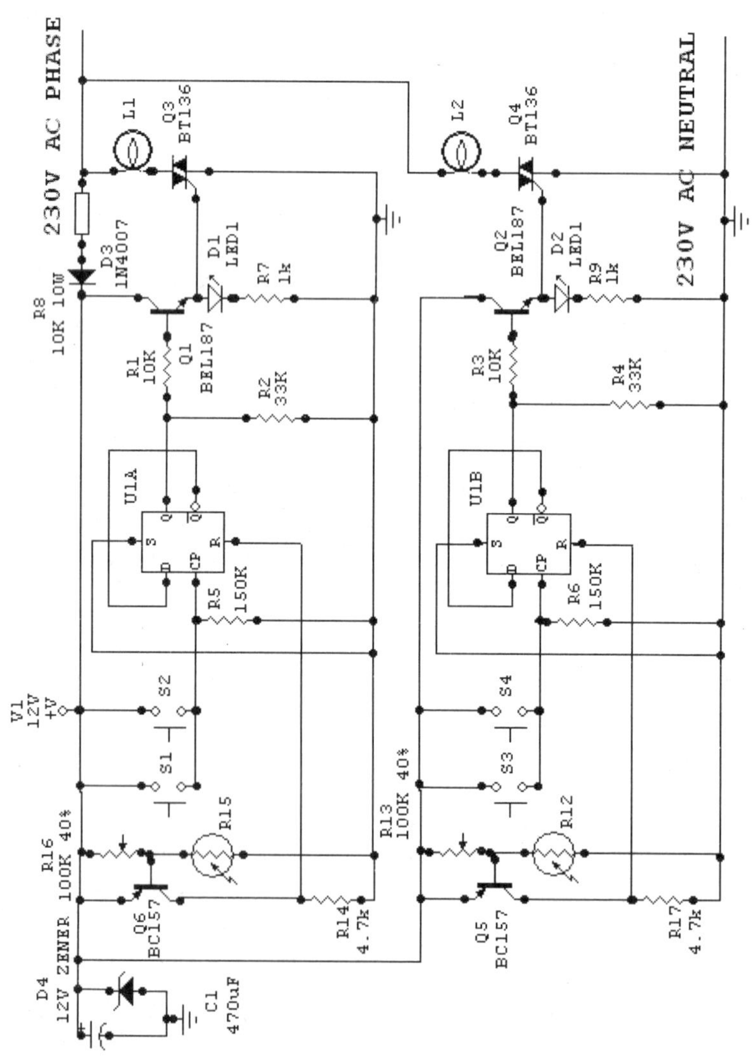

Schematic 25

Door Knob Alarm

Introduction

Here is circuit if you want an alarm to be sounded when somebody touches some thing. If any body touches say, the door handle, buzzer inside rings. But if want to keep the intruder away, keep the circuit on in the night or when you are away. It can be configured for use in a museum or display exhibition where you do not want visitors to touch the exhibits.

Description

Alarm sounds from speaker for 10 seconds after anybody touches the touch plates. This circuit using CD4011 can be used as a doorbell. The pin out and truth table are given ***Figure 32*** and ***Figure 33*** respectively.

PIN OUT

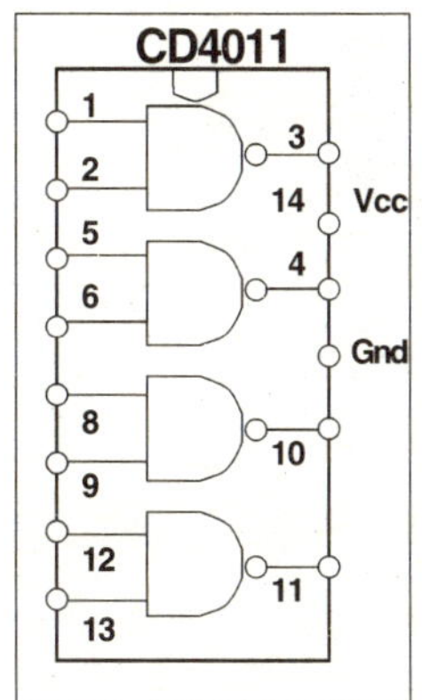

Figure 32

TRUTH TABLE

$Y = \overline{AB}$

Inputs		Output
A	B	Y
L	L	H
L	H	H
H	L	H
H	H	L

Figure 33

Gates are some of the fundamental building blocks of digital electronics. This CMOS IC, CD4011 has four two input NAND gates in one package. The truth table for this type of gate is given below. Inputs are taken as A and B. If both or any one of the inputs goes low, the output goes high. On the other hand, if both inputs go high, the output goes low. Converse happens in the case of AND gate. It can handle voltages from 3 to 15 volts.

Two NAND gates are used for touch sensing and other two gates are used for sounding alarm. Gates consisting of U_{1A} and U_{1B} act as a timer for 10 seconds which is activated by touching the touch probes. These can be conveniently concealed in a door handle or any other special object. If any body touches input gate at U_{1A} and U_{1B} it results in a high signal for 10 seconds at its output. The other two gates comprising of U_{1C} and U_{1D} make a high frequency oscillator. This is activated by a high signal at the output of U_{1A}. It is amplified by Darlington pair consisting of BC 147 and BEL187 which drives the speaker.

Construction

Construction is straightforward and this IC is a CMOS device. Construction on a Vero board should do well. A simple full wave supply of 12V would suffice, but do not go higher voltage than that. When these devices are operated near their upper limits, any small upward deviation can damage the IC. Take two small metallic touch plates and place them suitably at a place. Bridging them with fingers will activate the circuit.

Parts

Item	*No. reqd*	*Description*	*Designation*
1	1	10uF	C1
2	1	.01uF	C2
3	1	BEL187	Q1
4	1	BC147	Q2
5	4	100K	R1, R2,
6	1	10M	R3
7	2	10K	R4, R5
8	1	8 ohms	SPK1
9	2	CD4011	U1 A, B, C, D

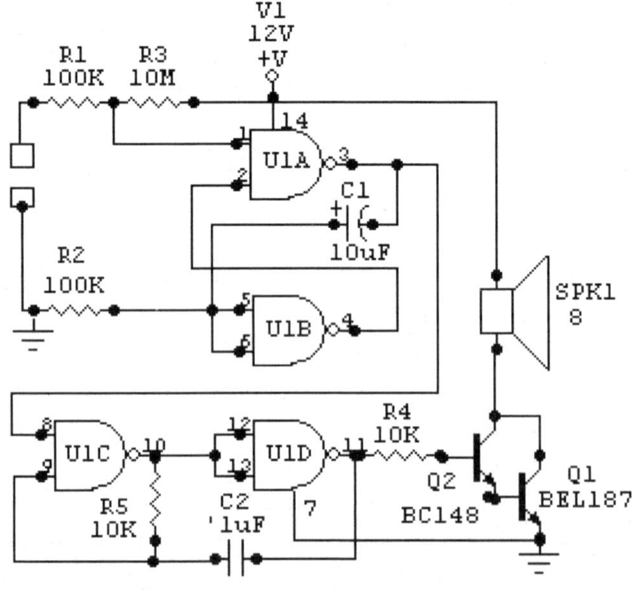

Schematic 26

Simple Quiz Circuit with SCR

Introduction

Here is a small and simple circuit to catch the first to punch the button in a quiz game. It is basically designed for two but can be expanded to accommodate more contestants. Present circuit works straightaway on mains but can be modified for use at low and less risky AC voltages also.

Description

Silicon Controlled Rectifier (SCR) is much similar to a conventional rectifier except that the conduction is controlled by a gate signal. SCR does not normally conduct until a signal applied at the gate. Once it is turned on, it continues to be in the ON state even after removal of the gate signal, as long as a minimum holding current is maintained or removed. The symbol of SCR is given in ***Figure 34.***

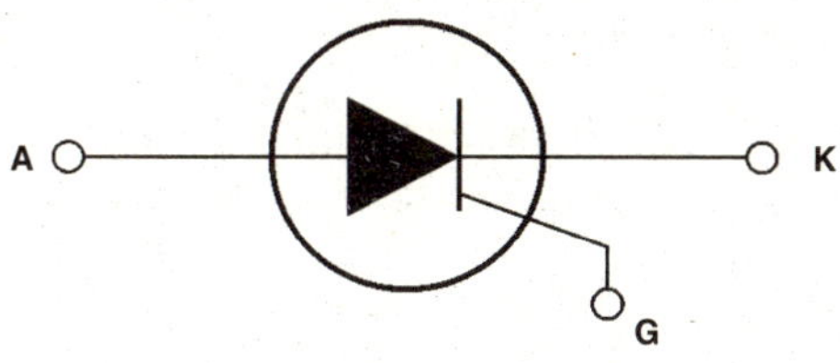

Figure 34

In a quiz game, we need to know, who pressed the switch first and a corresponding light bulb should glow indicating it or a buzzer should go on. Immediately all other switches should be disabled.

At the instance of power on, both SCRs will be in off state as both gates are at the ground level because of R1 and R2. Therefore, both the SCRs will not conduct, and both lamps will be off. Now let us say, if S1 is pressed first, the gate of corresponding SCR1 is pulled up, SCR1 conducts and latches on. L1 bulb glows and stays on because of SCR action. Now this SCR also makes diode D1 to conduct and pull down the gate of SCR2 disabling it. Hence it can not come on even if S2 is pulled up. Power up again to restart. Similar action is ensured when S2 is pressed. The circuit is shown in **Schematic 27.**

Construction

Present circuit is simple but it works directly on mains. If you are using a Veroboard, please keep adequate distance between the tracks such that arcing does not take place. Use mains operated bulbs.

Parts

Item	No. reqd	Description	Designation
1	2	1N4007	D1, D2
2	2	100W	L1, L2
3	2	5K	R1, R2
4	4	10K	R3, R4
5	2	47K	R5, R6
6	2	PUSH-TO-ON SWITCH	S1, S2
7	2	2N6241	SCR1, SCR2
		Or TYN604	
		Or TIC 106	
		or any other mains operable SCR with required current rating.	

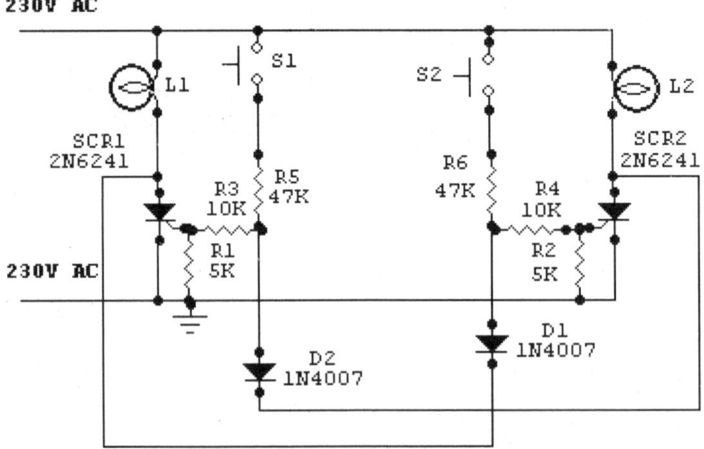

Schematic 27

Four Station Quiz

Introduction

Here we have another Quiz circuit using a number of CMOS ICs. Now the number of contestants can be easily increased even though the present circuit shows only four. It uses a two input Quad AND gate (74HCT08), a four input Dual AND gate (74HCT21), a dual D flip flop (CD4013), and a 555. Truth tables of the ICs are given here. If any one of the four contestants presses the switch first, LED corresponding to his station lights up and the alarm goes on. All other switches will be immediately disabled. The circuit has to be reset by a switch at the master to start the next question.

Description

The circuit is shown in **Schematic 28.** It consists of four ICs. 74HCT08 has four AND gates with two inputs each. 74HCT21 has two AND gates each with four inputs, of which only one AND gate is used in this circuit. The other can be used for expanding the number of contestants. CD4013 dual D type flip flop and two such ICs are used in the present circuit. There is one inevitable 7555 used as an astable multivibrator for alarm.

There are four push button switches, one for each contestant. Each switch is connected to one input of AND gate of 74HCT08. This input is made low by 1K resistance and can be pulled up by the push switch. All the other input gates are made common and connected to the output of 74 HCT21. Outputs of 74HCT08 are individually connected to the four set inputs of flip-flops in both CD4013s. Reset pins in CD4013 are made common, pulled down by 1K resistance and connected to reset switch. This switch is used to reset the circuit for the next question.

Now Q outputs of CD4013 are made common through diodes D1, D2, D3, and D4 and connected to the reset terminal of 7555. 7555 is wired as an astable multivibrator but its reset pin is pulled low by 3.3k resistor. Four LEDs D5, D6, D7, and D8 are connected at the Q outputs of CD 4013 to indicate which switch was pressed first. $\overline{Q}$ outputs are individually connected to the four inputs of one AND gate in 74HCT21.

Let us take the case of S1 contestant. As soon as he presses the switch, if he does so first, the output of his AND gate (U1A) goes high. This makes the SET pin of CD 4013 (U3A) high. Now output Q will be high and $\overline{Q}$ will be low. LED1 connected to Q will glow.

Now when Q goes high, it also brings this reset pin high through IN4148 diode and the alarm goes on. Now $\overline{Q}$ which is low is connected to an input gate of 74HCT21 (U2A) (It has four inputs). Out put of this IC is connected to all the second input gates of 7408. Hence they will all be pulled low. Now even if any other switches are pressed, they will not respond. There is a provision for an LED at the individual station taken from $\overline{Q}$ output to show which station has responded first. So there two sets of LEDs, one set at the master and the other at individual contestants.

All the Q outputs of both CD 4013 are connected to the reset pin of IC7555 through individual diodes (IN4148). *IN4148 diodes are fast recovery version, which look like small zener diodes.* Hence when any one of the Q outputs becomes high, alarm will be activated at 7555. All the reset pins of CD4013 are connected to a switch (S5) which acts as a master switch. To normalize the operations, reset should be pressed by the master. Then the quiz can start all over again. The fastest will now get the priority and all others will be disabled.

74HCT08

Pinout and truth table are given in ***Figure 35 and Figure 36*** respectively This IC has four AND gates with two inputs each in a single packing. As you can see from the truth table, output will be low if any or both of the inputs are low. Output goes high if only both inputs are high. It is an inverse of NAND gate already discussed.

PIN OUT

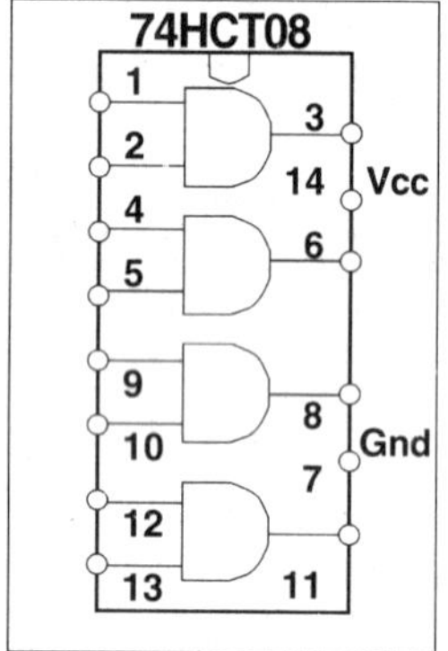

Figure 35

TRUTH TABLE

Inputs		Output
nA	nB	nA
L	L	L
L	H	L
H	L	H
H	H	H

Figure 36

74HCT21

Pinout and truth table are given in ***Figure 37 and Figure 38*** respectively This IC has two AND gates with four inputs each in a single packing. As you can see from the truth table, output will be low if any or all of the inputs are low. Output goes high if only all the inputs are high.

PIN OUT

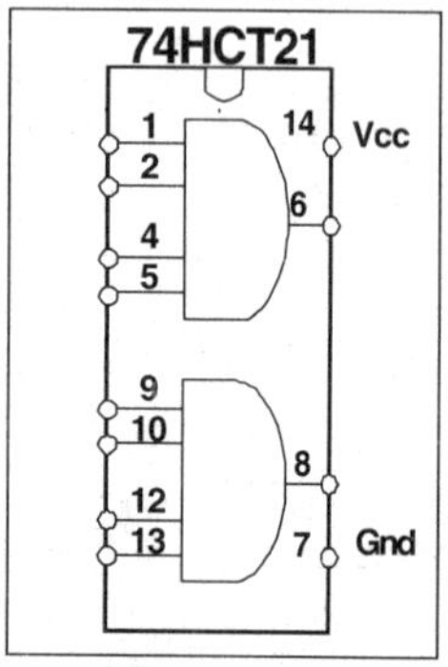

74HCT21

Figure 37

TRUTH TABLE

Input				Output
nA	nB	nC	nD	nY
L	X	X	X	L
X	L	X	X	L
X	X	L	X	L
X	X	X	L	L
H	H	H	H	H

Figure 38

Pin out and Truth Table of CD4013 ...Please See Earlier Circuit.

Construction

Now you have four ICs to deal with. Vero board will suffice, but think well for the best position of ICs so that a minimum number of interconnections are made. Switches must be rugged to stand the excitement of quiz competitions and should stand repetitive punches. You need to take three wires from the main console to individual contestant stations. One will be the power line and another is in the input line of AND gate, another is for the indication LED at the station. To save number of wires going to the stations this LED is given from the Q dashed output of the Flip-Flop. Power is regulated 5V, unless you can procure 74C versions of the ICs. Ground all unused inputs particularly of 74 HCT21.

Parts

Item	*No. reqd*	*Description*	*Designation*
1	1	250uF	C1
2	3	.1uF	C2, C3, C4
3	4	1N4148	D1, D2, D3, D4,
4	8	LED	D5, D6, D7, D8, D9, D10, D11, D12
5	14	1K	R1, R2, R3, R4, R5, R6, R7, R8, R9, R10, R11, R12, R13, R14
6	2	3.3K	R15, R16
7	5	Push to on switches	S1, S2, S3, S4, S5
8	1	8 Ohm Speaker	SPK1
9	1	74HCT08	U1
10	1	74LS21	U2
11	2	CD4013	U3, U4
12	1	555	U5

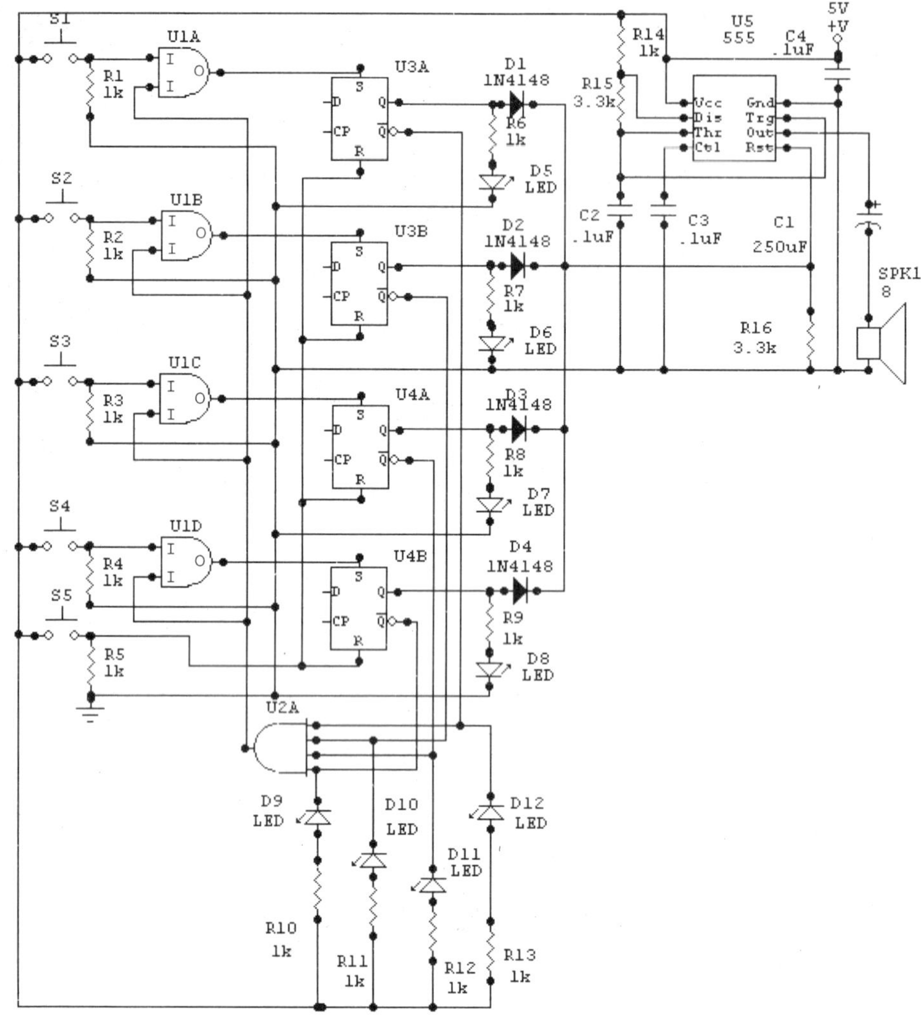

Schematic 28

Electronic Lock

Introduction

Here is an electronic code lock, which can be used as a door latch or key for ignition, etc. Operation is fairly tricky and there lies the beauty of the circuit. There are a number of switches, actually nine, which should be operated in a particular and owner - configured - sequence, or else the circuit will not drive the final SCR. The switches can of course be reconfigured as desired. SCR Output should be used with a suitable driver such as a solenoid for the door latch or a relay for the key.

Description

The circuit is shown in **Schematic 29.** CD 4066 quad bilateral switch is made use of here. 12 V DC powers the circuit through S1. External switches S2—S9 are not connected in the same order as their number and that is part of the trick.

S2 is a dummy switch, when pressed, LED D2 lights up only to fool the intruder. It is not connected to the rest of the circuit.

S3 is the next switch. This operates internal switch 1 of CD 4066. When this switch is pushed, it pulls up trigger terminal (Pin1), and switch across 13 and 2 (SW1) is closed. It stays closed because of the feedback action of 3.3M resistance (R1). D1 lights up indicating the closure of one switch in the sequence.

This powers the second internal switch (SW2) consisting of 5, 4, 3 pins. Power reaches Pin 5 and Pin 4 is the trigger terminal. When S5 switch is pushed on internal switch across 5 and 3 (SW2) closes. It charges C1 capacitor 47uf through 100K resistance (R3). It can now feed the next switch as long as the capacitor can hold charge. C1 is discharged through D3 and R5, which mean that next switch should be operated before this charge finishes.

To add to the confusion, the next switch is actually two switches in series comprising of S4 and S7 with trigger terminal at Pin 6. If they are pressed simultaneously, only if they are pressed simultaneously, internal switch across pins 8 and 9 (SW 3) closes. This charges 47uF capacitor (C2) through 100 k resistor (R6) which discharges through D4 and R7. Hence one has to press the next switch S8 before this charge is completed.

When S8 with trigger terminal at Pin 12 is operated in time, internal switch across pins 11 and 12 (SW4) closes.

SCR is fired now through R9. SCR drives a solenoid or a coil or any other drive mechanism of the lock. Final LED (D6) also lights up.

S9 is a blind switch only to fool the inadvertent user. S6 is another clever switch. This lights up LED D5 but also starts a piezo buzzer warning that somebody is fiddling with the lock. A 2200 uF capacitor charges and keeps the buzzer for some time. Use of capacitor is deliberate. It also makes the rogue user take a quick run.

Construction

Construction with CMOS IC is simple and straight. The trick here is to lay out the switches in a haphazard sequence, known only to the authorized user. Provision must also be made for easy change of code. With nine switches available, permutations are really many. Wiring must be carefully done to avoid false triggering.

Parts

Item	No. reqd	Description	Designation
1	1	Piezo Buzzer	BZ1
2	2	47uF	C1, C2
3	3	LED	D1, D2, D5
4	2	1N4148	D3, D4
5	1	1N4003	D6
6	3	3.3M	R1, R5, R7
7	3	1K	R2, R4, R8
8	2	100K	R3, R6
9	1	10K	R9
10	1	12V SOLINOID COIL	RLY2
11	8	Push to on switches	S2, S3, S4, S5, S6, S7, S8, S9,
12	1	SPST SWITCH	S1
13	1	SCR	SCR1
14	1	CD4066	SW1, SW2, SW3, SW4

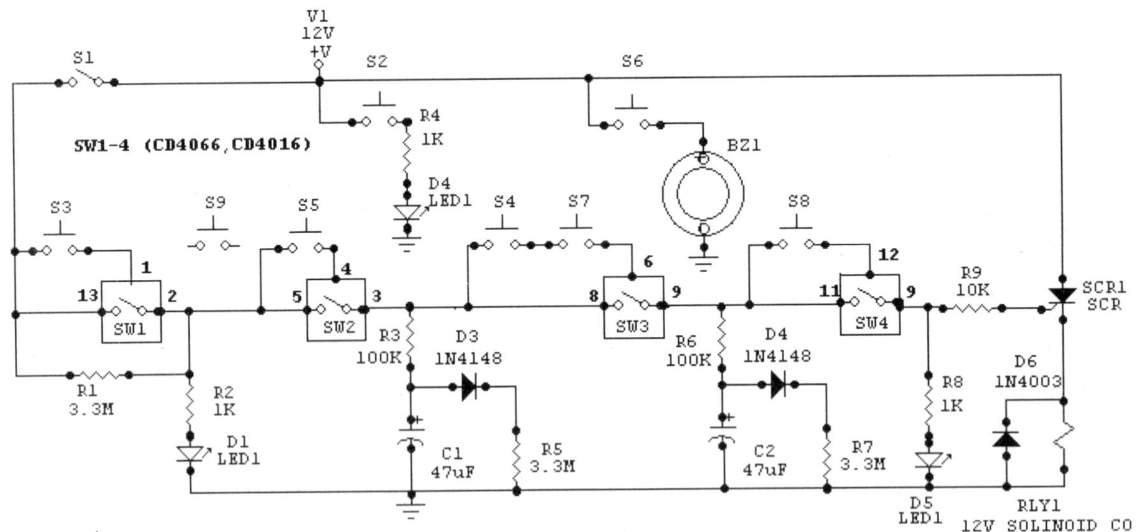

Schematic 29

Stitching Machine Motor Speed Control

Introduction

Motor operated stitching machines have a series of carbon buttons in an enclosure operated by foot pedal. As the pressure on the foot pedal is increased or decreased by the foot, these buttons come close or move farther away, their resistance changes and hence the speed of the motor. Even though crude, it seemed to be okay, until my wife complained.

She said that the speed is little too fast even at the minimum pressure on the foot pedal, particularly for some repair work or at embroidery. Most of the housewives also like a little more control over speed of the machine. So here you have it.

Description

The circuit is shown in **Schematic 30.** This is a standard triac speed control circuit much similar to domestic fan control circuit. Contrary to other triac circuits, you will find that an additional component known as diac is used in this circuit.

Triacs fire more symmetrically when used along with diac in AC power control applications. The diac is a bidirectional trigger diode which does not conduct (except for a small leakage current) until the break over voltage is reached. Its function is designed specifically to trigger a triac or SCR.

In the beginning triac Q1 is not conducting; C1 is charged through variable resistor R3. This charge is coupled to Diac through R1, R3. When trigger level of the diac is reached (about 36 V), D1 fires and Triac TR1 is switched on. R3 and C1 combination sets the firing point of the triac from zero crossing along with R1 and R2. L1 and C3 combination acts radio frequency filter for the radio interference caused by triac firing.

Construction

Entire circuit operates on mains. Care must be exercised when mounting components on normal Veroboard is risky. Remove alternate tracks and mount components. Triac should be mounted on a small heat sink as the triac tends to get hot particularly at lower speeds. All capacitors are polyester or polycarbonate rated at 600V or more. R1 is a preset for minimum speed control. Adjust this according to your requirement. R2 is the linear variable resistor like the volume control in the radios. Use the one with plastic shaft. L1 is a radio interference choke. Take 28 gage winding wire and make 10 turns on a 6 mm former. It can be wound on a round capacitor for C3 and even one end can be soldered to it also.

Parts

Item	No. reqd	Description	Designation
1	3	.1uF/600v	C1, C2, C3
2	1	Diac	D1
3	1	Coil	L1
4	1	MOTOR	M1
5	1	BT136	Q1
6	1	47K	R3
7	1	47K variable	R1
8	1	470K variable	R2

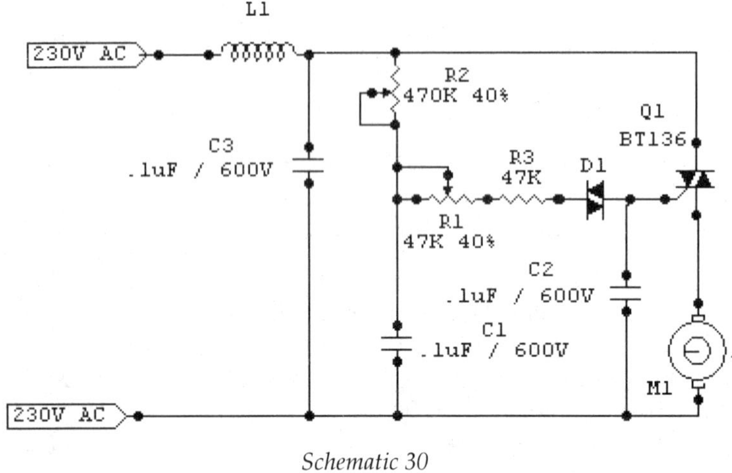

Schematic 30

Telephone Light

Introduction

Telephone rings in the dead of the night and you grope in darkness for the light switch or may be for the cordless handset. Here is ingenious circuit which makes a light bulb glow whenever the telephone rings.

Description

This is a good application where mains are triggered by triac by the action of an opto-coupler. The circuit is shown in **Schematic 31.**

Opto-coupler is device where the signal is coupled from input to the out put optically. It offers full isolation for both sides. Usually a LED is placed on one side internally and a photo transistor on the other, all housed in an opaque package similar to an IC. Unlike a transformer, opto coupler also allows DC linking. MCT2E opto coupler consists of an infrared LED (Pins 1, 2) driving a silicon phototransistor (Pins 4, 5, 6) in a 6 pin DIL package. Pinout details are given in ***Figure 39.***

PIN OUT

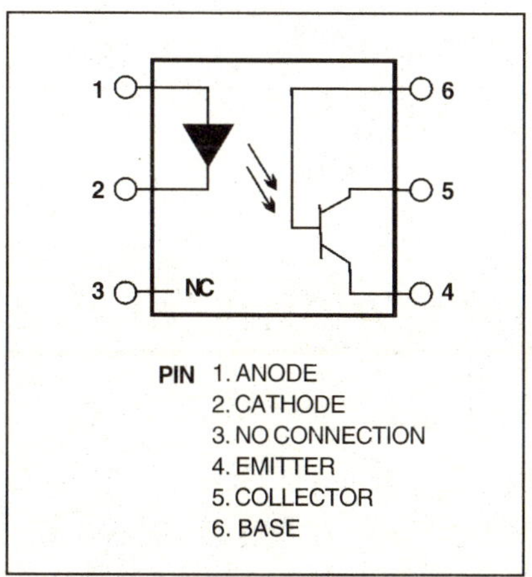

Figure 39

IR LED across pins 1 and 2 in MC2E opto-coupler is connected to the telephone line with a 47 k resistor. When the telephone rings, LED lights up internally. It is sensed by the phototransistor across pins 4, 5 and 6. This drives the gate and the triac fires making the bulb glow.

Construction

You are warned that the circuit operates mains. There are only few components to be mounted. Check the pin-out of MCT2E and fix it properly. See that no arcing develops across the tracks or components. Use of optocoupler here, effectively isolates it from mains

Parts

Item	*No. reqd*	*Description*	*Designation*
1	1	100W	L1
2	1	BT136	Q1
3	1	1K	R1
4	1	47K	R2
5	1	MC2E	U1

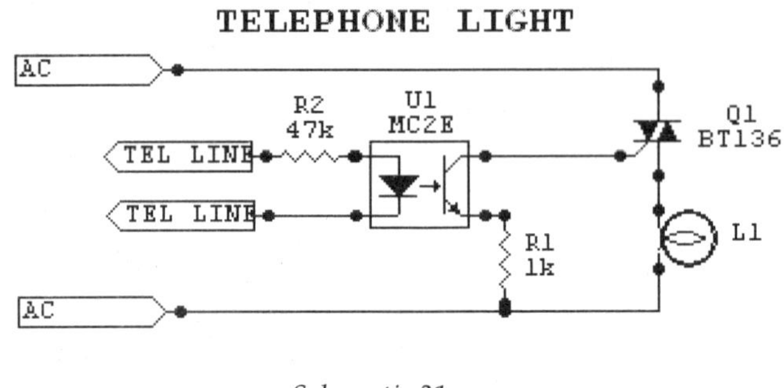

Schematic 31

Morning Alarm for Babies

Introduction

Here is a pleasant wake up alarm for the babies. As the morning light breaks in, the alarm gently wakes up the baby with a soothing musical note. What a pleasing way to make the baby rise to the day!

Description

As already described, LDR has a very high resistance in darkness, which falls low as the light falls on it. Hence Q1 does not conduct in darkness, as it is reverse biased by VR1. As the daylight breaks Q1 goes into conduction and powers UM66 musical IC. Musical note at its output is amplified by Q2 and is fed into the 8-ohm speaker.

There a number of UM66 T chips, each giving a different note. You may also add a suitable voice COB like a sloka or a prayer in its place. The circuit is shown in **Schematic 32.**

Construction

You may use 3V with dry cells as the power supply, but operation at 3.6 V is good. It is better to use three or four Ni Cad or Ni Mh cells as they can be recharged and reused. Straight forward construction on a Vero board is all needed. Use plastic battery box available in the market to fix the batteries. Adjust 220k preset to the desired sensitivity, thereby at the required intensity of daylight, when the alarm should start singing.

Parts

Item	No. reqd	Description	Designation
1	2	3V	DRY CELLS
		OR	
3	3.6V	Ni.Mh CELLS	
2	1	LDR	R4
3	4	BEL187	Q1, Q2
4	4	1K	R1
5	3	470	R2
6	2	220K variable	R3
7	1	8 Ohms	SPK1

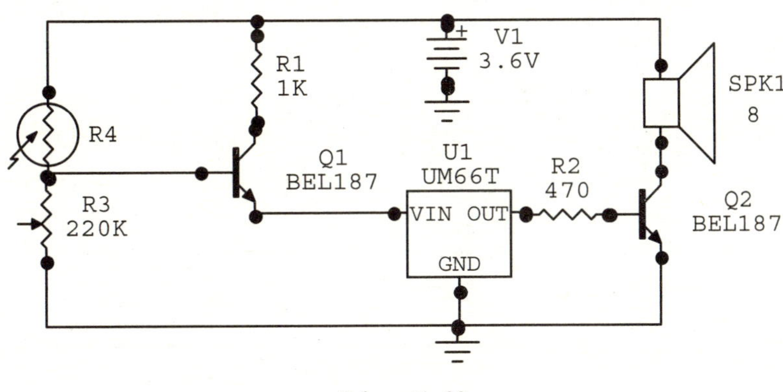

Schematic 32

White LED Night Light

Introduction

A better choice of a night light would be to use white LEDs which have become quite cheap and affordable of late. Circuit here works on mains and it can be fixed in small enclosures.

Description

0.22 mFd /600V capacitor drops AC voltage to about 15 volts. It is rectified by mains bridge (4No.s of IN4007 diodes will do well) and filtered by a 100 mFd capacitor. Zener diode limits the circuit at 15 volts. A series of 4 LEDs are connected now along with a 220 ohms current limiting resistor. You may increase the value of this resistor to reduce the current through

LEDs. Another series of LEDs can be added in parallel to it to increase brightness. White LEDs take about 3.6 volts for operation compared to 1.8 Volts for red LED and 2.0 Volts for green, orange and yellow LEDs. The circuit is shown in **Schematic 33.**

Construction

You may fix all the LEDs in a transparent enclosure and directly mount it on the wall plug. Even though the voltage is brought down by the capacitor, the circuit is not isolated and you will get a nasty shock by touching it while in operation. 230V mains supply is connected through .22 capacitor and 220 ohms resistor.

Parts

Item	*No. reqd*	*Description*	*Designation*
1	1	D1	MAINS BRIDGE
		OR	
	4	DIODES	IN4007
2	4	D2, D3, D4, D5	WHITE LEDs
3	1	D6	15V ZENER
4	1	C1	0.22 / 600V or 0.22 / 275V ac
5	1	C2	100 mFd / 25V
6	3	R1	220 OHMS

WHITE LED NIGHT LIGHT

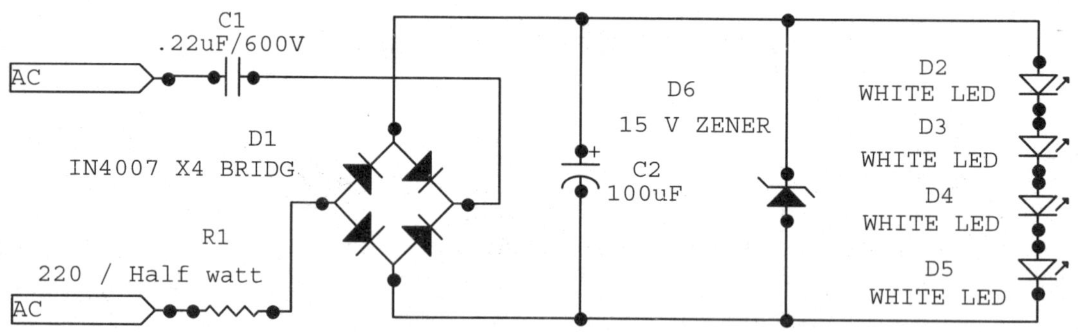

Schematic 33

Chandelier Dimmer

Introduction

Many of the chandeliers have a pull chord switch. Now add a novelty to your chandelier and surprise the guests. Place a small wire down the lamps from this circuit and touch it once, they will come on dimly. Touch again they glow with medium brightness. Another touch drives them full. One more touch the lights will be fully off.

Description

Here is an attractive application of Touch Dimmer IC, TT6061, an 8 pin CMOS IC which has three steps of touch dimming for incandescent lamps. Because of its high sensitivity, a long wire can be connected to the touch sensor. The circuit has minimum external components. Regulated supply of 6.2V is directly taken from the mains with an operating current of 1mA through R1, D2, D1, and C1. R1 drops the mains voltage, D2 rectifies it, C1 filters it and D1 regulates it at 6.2 V. Pin out and description of pins is given below.

Just a copper plate of 1cm X 1cm can be the touch plate or even the end of the lead wire will work well. AC hum pervades all over and it is sensed from this plate by the touch detector through two No.s of 1kpf / 2kV capacitors in series (C2, and C3). It is then registered in a step register and connected to the counter / decoder. Line frequency signal is fed into the IC through R2 to Pin No.2. At the zero crossing, the triac (BT136) is triggered with appropriate pulse, which can drive up to 500W with a suitable heat sink. The circuit is shown in **Schematic 34.**

PIN OUT (Top View)

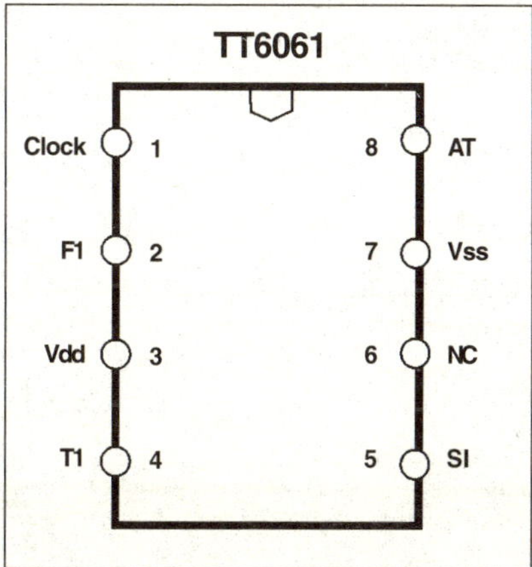

Figure 40

Pin Description

1. CK CLOCK INPUT
2. FI LINE FREQUENCY 50/60Hz
3. VDD POWER INPUT
4. TI TOUCH INPUT
5. SI SENSOR CONTROL INPUT
6. NC
7. VSS POWER VSS
8. AT OUTPUT

Do not replace two series capacitors C2, C3, with a single capacitor or with a capacitor of a lesser voltage rating. They are connected between touch input pin (4) and the touch plate to remove the shock potential from the touch plate.

Construction

Mains potential exists in the circuit and you are warned. Take all the precautions which you take while making or working on mains circuits. Place a small heat sink on the triacs. Take a note of the power consumed by the chandelier. If it is more than the capability of this triac, use a higher rated one.

Parts

Item	*No. reqd*	*Description*	*Designation*
1	1	100uF	C1
2	2	1kpF/2KV	C2, C3
3	1	.047uF	C4
4	1	.47uF	C5
5	1	6.2V Zener	D1
6	1	1N4007	D2
7	2	1N4148	D3, D4
8	1	100W	L1
9	1	BT136	Q1
10	1	39k/2W	R1
11	1	1.5M	R2
12	1	820K	R3
13	1	1K	R4
14	1	6.8M	R5
15	1	10K	R6
16	1	TT6061	U1

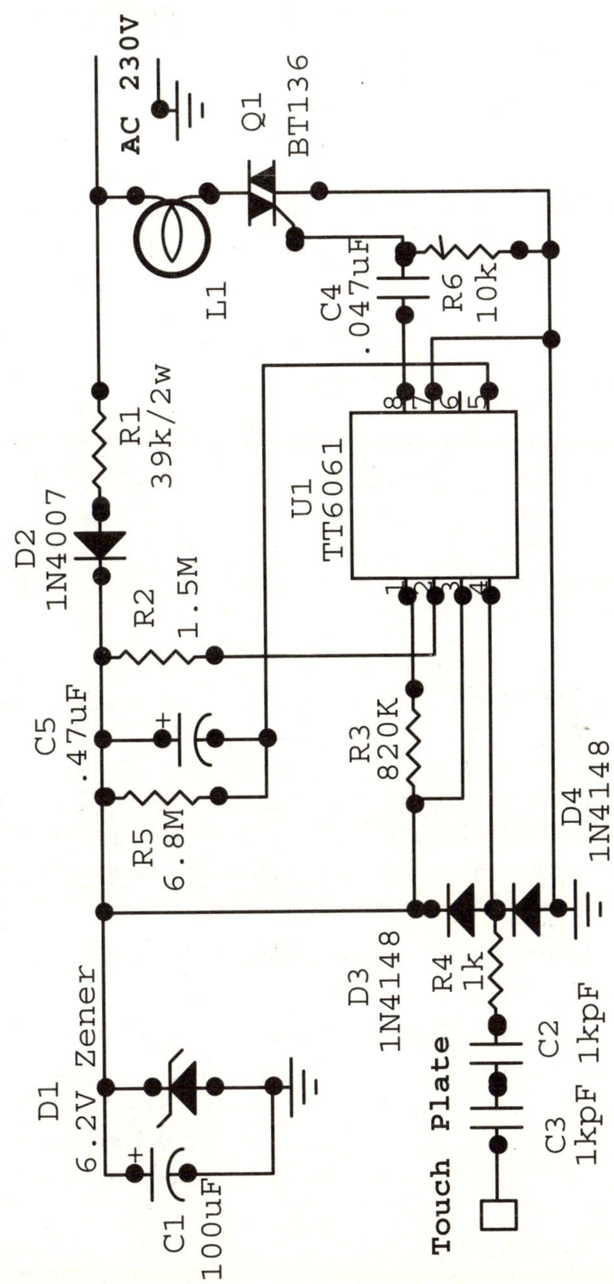

Schematic 34

Audio Circuits

Here we have a series of audio amplifiers circuit starting from a simple 1W to powers reaching 20 watts. These circuits start with a simple condenser microphone adopter (Shall we say?) and build up to amplifiers of several watts.

Do you want more power?

What is watt?

There is a general confusion prevailing in the audio industry each one claiming more output power than the other. Peak Music Power outputs powers of more than 7000 watts are claimed. These ratings are baffling to the general public as the salesmen normally contend and contest that it is that actual power their systems handle. Half baked sales men do talk as they like but it is unfortunate that the reputed manufacturers fell into this bandwagon. Public are falling head downwards with a kind of a personal prestige value, as usual with false figures.

A simple mathematical calculation shows that to achieve 7000 watts of power with 12 V power supply would need approximately 583 amperes. Think of the transformer, rectifiers and other components to handle this power. When one opens insides of the gadgets, takes one look at the ICs used, it will be surprising that the actual RMS powers fall into single digits. A television in a normal living room is used with less than 1 W of audio power. I laughed at the claims of audio system manufacturers who rated their systems at about 400W of PMPO, where the IC inside can hardly drive 6W. I am sure they would never be able to prove these hefty power ratings even in their laboratory conditions, let alone in normal domestic circumstances. I contend that we live in wilderness.

Please also note that simply by using higher wattage speakers, your out puts will not increase. As a matter of fact, delivered acoustic power may drop appreciably.

Read a small tutorial in the end about audio amplifiers and trouble shooting. Obviously, the following versions give only true power at RMS. Hence the power may not sound high. Nevertheless there are a number of circuits to choose from. Select one depending on your personal requirements and capacity. We will start with a small microphone, preamplifier and advance to more and more powerful amplifiers with some useful applications.

A Collar Mike

Introduction

Microphone is added here for the very use it has, and by the ease of making one. Cost of a good microphone made like this would be less than Rs.50, even if 10 meters of Mike cable is used. It can be used as a good quality microphone. Here is a simple circuit that makes the condenser mikes work, which can also be used as a collar mike. Condenser mikes are rugged, have excellent frequency response, and they develop fairly high signal output. They can certainly be implemented to test later circuits. Make this and I am sure you will make many for many of your friends.

Description

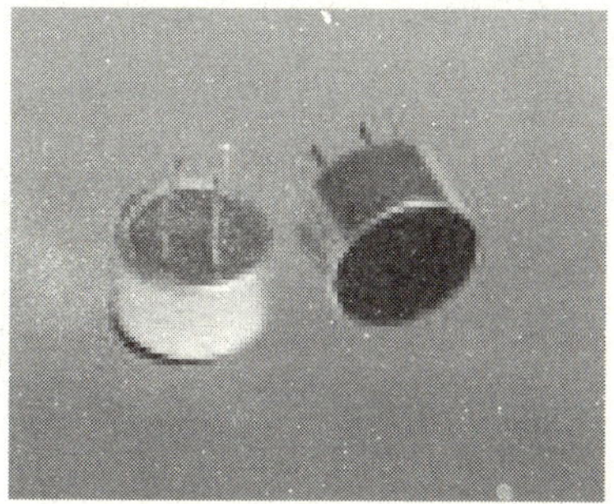

Figure 41

These ubiquitous devices are seen everywhere, in portable tape recorders to cell phones. These are small, sensitive, and rugged and have an excellent frequency response, and high signal to noise ratio. They deliver higher level of signal into preamplifier. It has an internal field effect transistor which needs power at less than half a milliampere. Condenser mike has two terminals; one terminal can be seen as connected to the body of the mike. The other terminal is the HOT terminal. It doubles up as an output terminal and power +ve terminal. A single pen cell can power it.

Circuit is shown in **Schematic 35.** Condenser mike is biased with a 3.3K resistor to power the internal FET. Output is also taken from this terminal to the Mike (MIC) input of an amplifier. There must be a capacitor at the input of the preamplifier to block the DC voltage from condenser mike. Preamplifiers generally have this capacitor at their input. Negative terminal of the battery is connected to the ground terminal of the mike. It is also connected to the ground of the amplifier through the shield of the connecting mike cable.

Construction

Connect 3.3 K resistor to the positive side of single cell battery holder with the other end of the resistor to the HOT terminal through a mike cable. Negative side of the battery holder is connected to the ground terminal of Mike. Solder a shield wire of required length to the terminals of the mike. Central core wire goes to the hot or output terminal and the external shield goes to the ground terminal. The other end of the shield wire is connected to the appropriate connector suitable to the power amplifier. Use good quality shield wire and appropriate input jack pin suitable for the amplifier. Make two such mikes for stereo applications

Note

I often advised people to use a used battery cell (not a dead cell) to power up collar mikes. It is necessary to keep the mike faraway from the speakers to avoid annoying feedback (howling). If higher voltages are used to power these mikes, either by batteries or otherwise, say in a circuit, you have to change the biasing resistor correspondingly.

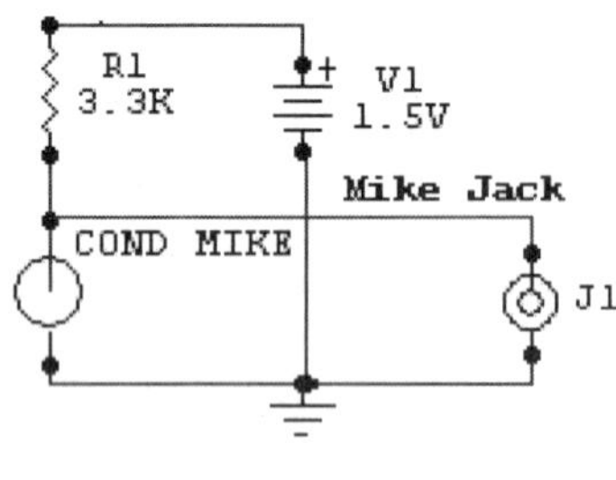

Schematic 35

Preamplifier with LA3161

Introduction

Preamplifiers are used to amplify low level signals such as those from mikes, tape heads before they are fed into power amplifiers. Power amplifiers are generally less sensitive. Frequency response also can be suitably trimmed and modified at preamp stage. LA 3161 is one of those widely used in tape decks and amplifiers as a stereo preamplifier.

Description

Block Diagram is shown in ***Figure 42.*** *LA 3161 has two low noise preamplifiers with good ripple rejection on chip catering to stereo applications. External part count is low and Single In line (SIL))(* ***Figure 43)*** *package makes mounting easy. While the operating voltage is 9V, the IC can tolerate voltages up to 18V. Typical input resistance is 100K and output resistance is 10K with an open loop gain of 78dB. Block diagram of the IC is given below. Input is given at Pin 1 and 8, output is taken at Pin 3 and 6, and negative Feedback is given at Pin 2 and 7. Power is at Pin 5 and Pin 4 is the ground terminal. There is an internal voltage regulator.* Circuit is shown in **Schematic 37.**

BLOCK DIAGRAM

LA3161

Input 1
NF 2
3 Output
4 Vcc
Input 8
NF 7
6 Output
5 Gnd

Figure 42

Single in Line (SIL) Package

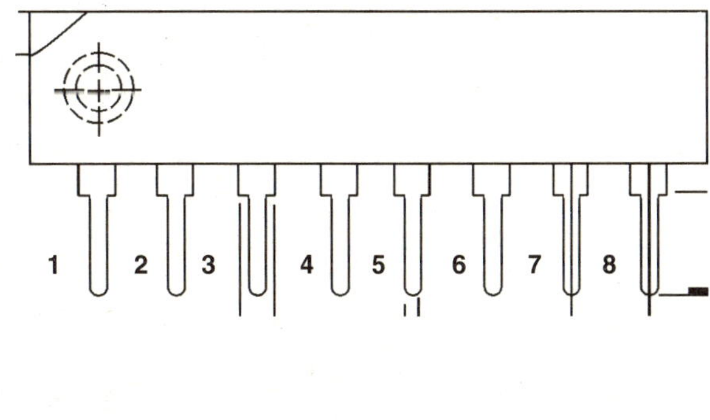

Figure 43

Construction

Ready made PCBs, even populated PCBs are available using this very useful IC. You can still build one, with a Veroboard provided proper care is taken about the ground returns. *It simply means that one should not connect ground terminals of output and input at the same place.* This will create serious oscillations and normal hobbyist will be left in the woods. Please read general instructions for working with amplifiers in the end.

Parts

Item	*No. reqd*	*Description*	*Designation*
1	2	10uF	C1, C2, C7, C8
2	2	47uF	C3, C6, C9
3	2	0.033uF	C4, C5
4	2	1000pf	C10, C11
5	4	100K	R1, R3, R10, R12
6	2	22	R2, R11
7	2	3.3K	R4, R9
8	2	2.2K	R5, R8
9	2	100Kvariable	R6, R7
10	1	150	C13
11	1	LA3161	U1, U2

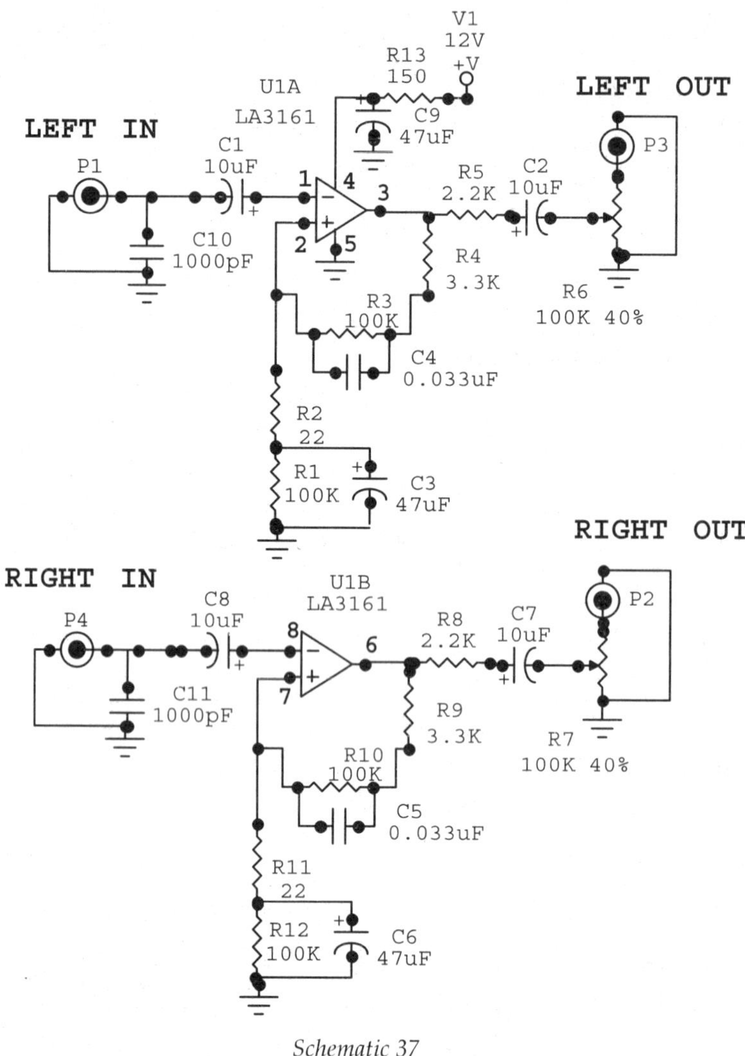

Schematic 37

LM 386 Audio Amplifier

Introduction

LM386 is one of the earliest audio ICs but is still going strong. It works with a few external components and is indispensable for general hobby circuits where audio amplifier of low power is needed. It is mini audio power amplifier which would prove essential as a test amplifier for solving typical problems in the audio circuits of TVs, telephones, etc. It can be used as an output driver for certain tone and oscillator circuits.

The following circuit is useful as a test mike but can be adopted for any of those requirements.

Supply voltage can range from 4 to 12 V. Gain is internally set to 20 but it can be increased to any value up to 200 by a manipulation of external capacitor and resistor between Pin 1 and 8. Quiescent current drain is only 24 milliwatts at 6V operation. LM 386 N-4 can operate up to 16V and can deliver 1W of power into 32 ohm speakers. With these features LM386 is a superb startup amplifier generally used as an output device for a voice IC or radio receiver. You should make two such amplifiers to make it stereo. External line in can be connected at C4 after switching out the mike.

Description

LM 386 has an internally fixed gain of 20 as in the case of the present circuit. It can be increased to 200 by bypassing 1K resistor by switch S1. With resistor in place the gain is 50 and without both capacitor (C2) and resistor (R3), (i.e., by keeping both pins 1 and 8 open) the gain will be 20. These components actually bypass an internal resistor of 1.2 K across pins 1 and 8. .047uF capacitor (C5) and 10K ohms resistor between pin 5 and pin 1 will boost the bass frequencies and are optional. Pin out of the IC is given in ***Figure 44.***

PIN OUT

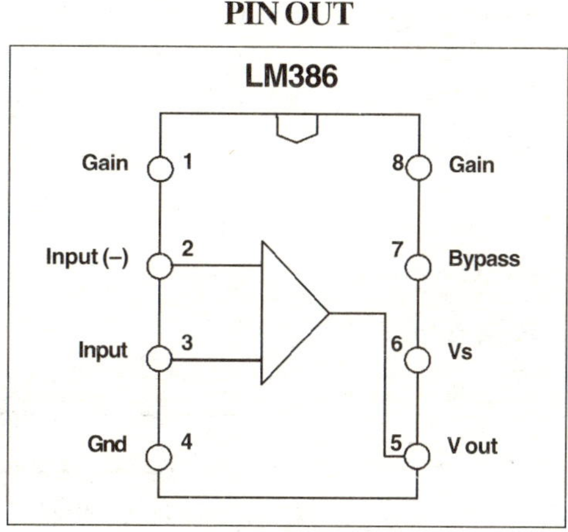

Figure 44

Circuit is shown in **Schematic 38.** Electret mike is powered by 10K resistor (R4) and is coupled to the amplifier by .47 uF capacitor (C4). R1 is volume control. Input is given at Pin 3 in the circuit and output is taken at Pin 5 through a capacitor (C1) of 470 uF to a speaker of 8 ohms. There is an additional Bass boost circuit given along side the main circuit which can be connected to boost low frequencies.

Construction

The circuit may be built on a Veroboard provided the precautions suggested in the earlier circuit are taken care of. Hence precaution must also be taken to avoid motor boating due to bad power supply filtering at the ICs.

Note

As a general rule do not use speakers of less impedance than prescribed in the datasheets. Use of such speakers will increase the output power more than designed and the IC may pack up.

Parts

Item	No. reqd	Description	Designation
1	1	470uF	C1
2	1	10uF	C2
3	1	1uF	C3
4	1	.47uF	C4
5	1	.047uF	C5
6	1	10K	R1
7	1	1	R2
8	1	1K	R3
9	2	10K	R4, R5
10	1	SWITCH	S1
11	1	8 Ohms	SPK1
12	1	LM386	U1

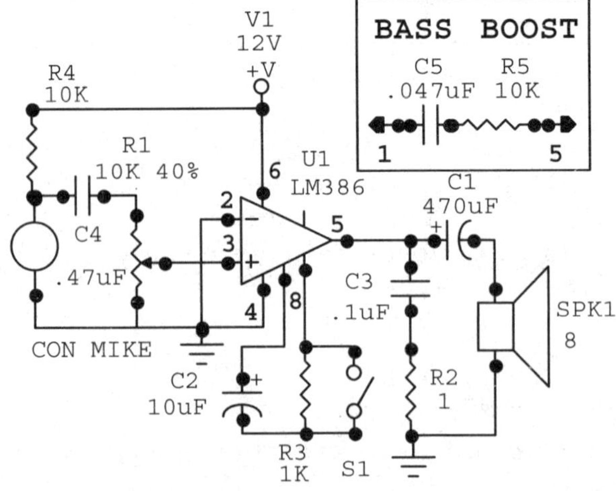

Schematic 38

Door Phone

Introduction

LM 386 is just right for a door phone application. It is very useful application in the present urban households. One condenser microphone and a speaker are kept conveniently outside and another set with amplifier are kept inside. With a door phone installed at the main door, it will be easy to recognize the caller outside the door. Nevertheless, it is still a better method for shooing away those indomitable salesmen with bodybuilder or body toner, or some other plastic gadget.

Description

Circuit is shown in **Schematic 39.** The amplifier along with the power supply is kept inside fully accessible by the insider. One microphone and speaker are kept outside the main door such that the caller can speak into the mike and listen to the speaker. When the caller knocks the door, insider can switch on the door phone and ask him to identify. Then the insider can switch on the outside mike to allow him to speak. This activates the inside speaker but switches off the inside mike. When the insider releases this switch, inside mike and outside speaker will be connected. This is simplex system.

Construction

There are two condenser microphones (MIC1, MIC2) biased by R3 (10k resistor). Bias is switched by S1 such that any one mike will operate. Mikes are coupled to the amplifier by C2, C7. VR1 is the volume control. Similarly there are two speakers (SPK1, SPK2) which are switched by S2 from the amplifier. Both these switches are operated by the insider and are ganged. The switches are ganged so that when outside mike is on, inside speaker will be connected and vice versa. R1 preset is a gain control, which you can adjust and leave it at optimum level. You still have volume control on hand. Please note that all these controls should be inside in the hands of the householder.

This amplifier is used with a power supply of 12V full wave. Regulated supply is not required. Use good quality capacitors as bad filtering would result in annoying hum.

Note

Read tips and tricks at the end to know more about audio amplifiers. Use shield wire for all the interconnections from mikes, switches and volume control to amplifier. Do not terminate ground connections as you like.

Parts

Item	No. reqd	Description	Designation
1	1	.10uF	C1
2	2	.47uF	C2, C7
3	1	.1uF	C3
4	1	470uF	C4
5	1	100uF	C5
6	1	1K preset	R1
7	1	1	R2
8	1	10K	R3
9	1	1K	R4
10	2	10K volume control	VR1
11	1	SPDT SWITCH	S1, S2 (Ganged)
12	1	8 Ohms	SPK1, SPK2
13	1	LM386	U1
14	2	COND.MIKE	Mike1, Mike2

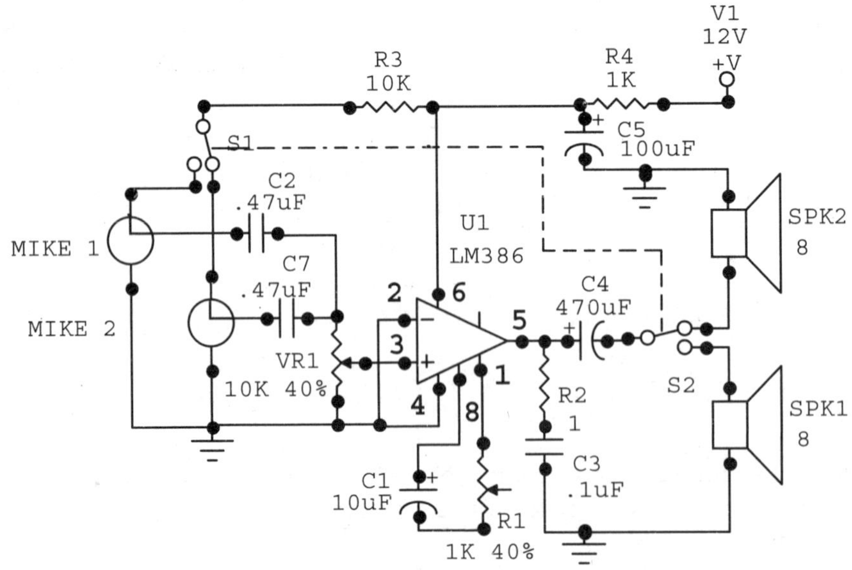

Schematic 39

LA4555 Audio Amplifier

Introduction

LA 4555 is basically a stereo amplifier with 2.3 watts into 4 ohms speakers at a total distortion of 10%. With a bridge circuit, it can be configured as mono amplifier delivering 4.6 watts. It has an input impedance of 30K and gain of 51 dB. It has an excellent voltage range of 3 to 13 volts. Both mono and stereo circuits are shown here.

Description

Pin out is given in ***Figure* 45.** Input is given at Pin 8 in the mono circuit and output is taken at Pin 11 through a capacitor (C7) of 470 uF to a speaker of 4 ohms. In the case of stereo circuit, input is given at 5 and 8 pins and output is taken out at 2 and 11 pins respectively for left and right channels through the blocking capacitors C4 and C7.

A detailed description of the pin out will help now and in the future.

In the stereo circuit,

C5, C2 are feedback capacitors, which dictate the lower cut off frequency.
C1, C6 are bootstrap capacitors. If the capacitor value is reduced from the recommended 47uF, output at flow frequencies falls.
C11, C10 are oscillation blocking capacitors. Polyester film capacitors are preferable.
C7, C4 are output coupling capacitors. Lower cutoff frequency depends on their value and quality.
C3 is the ripple filter or decoupling capacitor.
C8, C9 are the power source capacitors.
R2, R2 are oscillation blocking resistors.

ICs dissipate heat as they dissipate more and more power at more and more voltages. The heat must be removed continuously such that IC operates at rated temperature. Failure to do so will result in thermal runaway. If the IC is well protected, output power will fall to a safe area. If not it will eventually fail and pack up. Copper foil area is made as large as possible in the vicinity of IC to dissipate more heat.

*Then heat sink, thermally conductive material such as copper or aluminum is mounted on IC to remove the heat as it develops. It must be of enough size. In case of LA4555, the IC has fins which are soldered to the PCB and a small heat sink also can be soldered along with it. Solder copper heat sink as shown in the figure below. Aluminum cannot be easily soldered. Method of mounting heat sink is shown in **Figure 46.***

Stereo circuit is given in **Schematic 40** and mono circuit in **Schematic 41** respectively.

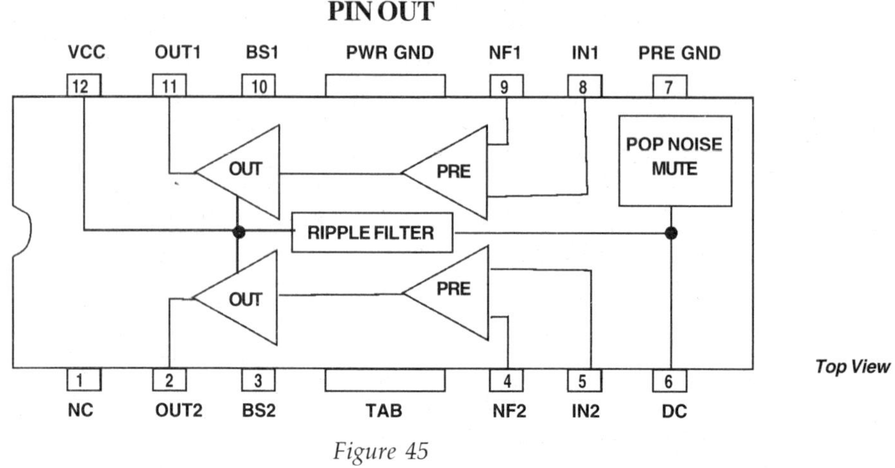

Figure 45

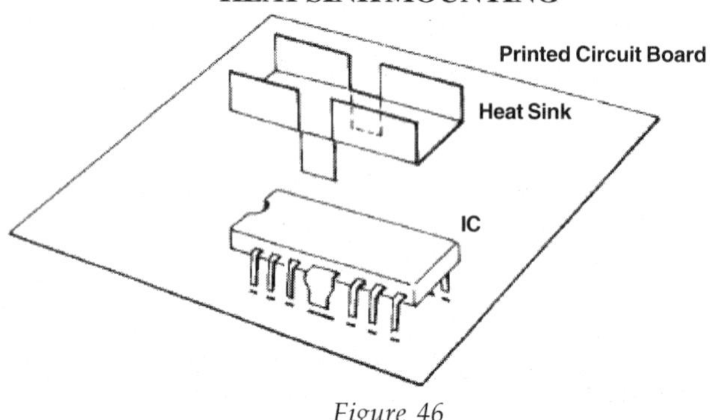

Figure 46

Construction

If you are using Veroboard, you may have to cut slots to fit the fins in it. Use heat sink as suggested above. *Please read about audio problems at the end of the book and follow the precautions.*

Parts

Item	*No. reqd*	*Description*	*Designation*
1	5	100uF	C1, C2, C3, C5, C6
2	3	470uF	C4, C7, C8
3	3	.1uF	C9, C10, C11
4	2	1R	R1, R2
5	2	8	SPK1, SPK2
6	1	IC	LA4555

For mono only one speaker will be required

STEREO CIRCUIT

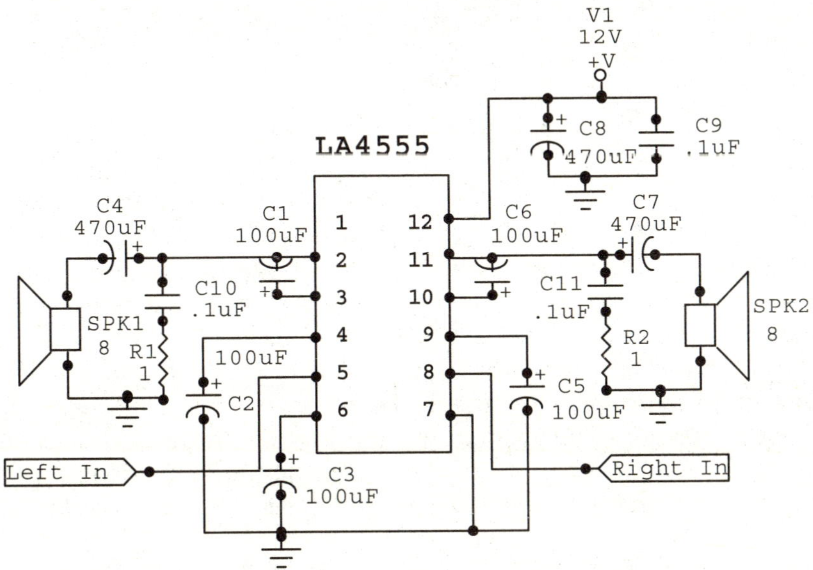

Schematic 40

MONO CIRCUIT

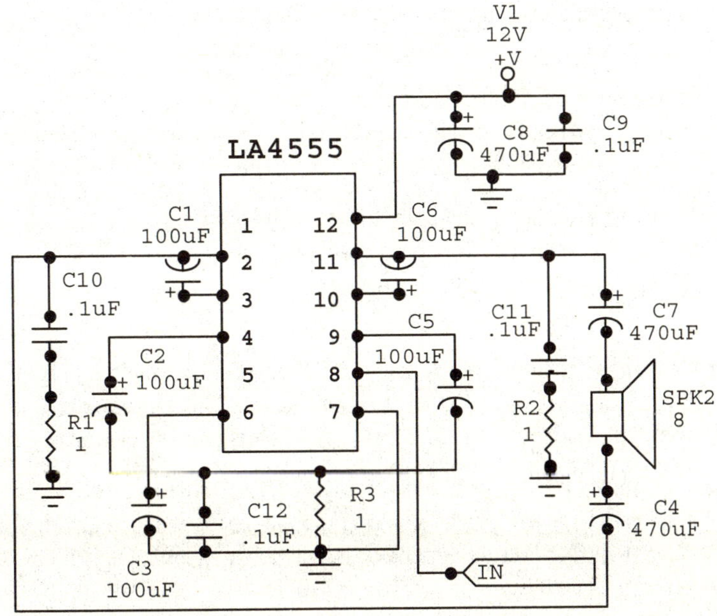

Schematic 41

Full Duplex Intercom with KA6283

Introduction

Here is an ingenious use of stereo application of CD6283 or KA6283. It is one of those widely used audio ICs now which delivers about 1W at 6V and 2.5 W at 9V in stereo loads of 4 ohms. This circuit is shown as an example for door phone or intercom, but you may use it as a simple stereo amplifier for any other purpose. Left and right channels of stereo amplifiers are made use of to power mikes and speakers of two intercom stations. This circuit can be used as full duplex intercom between two parties, say boss and the secretary, or main household and the security at the gate or several floors below. It is very useful as door phone as described in the earlier circuit but now we do not have any switches. It allows a hands - free – operation which permits multi tasking by the users. There is no push - to - talk button as in the conventional sets. I was using this system for communication when we were living in a two storied building. Housed in a SIL package, it has few external components, (only two resistors and other capacitors, it is easy to assemble and the heat sink is easy to fix.

Description

IC has excellent features like wide supply range of 6 to 15 V, thermal shutdown and minimum external parts count. IC can deliver 1W at 6V and 2.5W at 9V and 4.5W at 12V into 4 ohms speakers. While the schematic shows application of duplex intercom, IC part can alone be used for any other audio application using inputs at pins 5 and 7. Q1 and Q2 act as mike amplifiers and IC as the main amplifier.

Circuit is given in **Schematic 42.** In this schematic, left channel mike is kept at station1 but its speaker is kept at station 2. Mike from right channel is kept at station 2 but its speaker is kept at station 1.

Mike1 is powered by 6.8k resistor (R5) and is amplified by Q1 low noise transistor BC149. Volume level is adjusted by VR1 and fed into the left amplifier of CD6283. Its output is taken out to Speaker 2 kept at another place. The master speaker can now speak to the slave or allow him to speak into the mike.

Mike 2 signal is powered by another 6.8 K (R6) resistor and is amplified by Q2. Right channel amplifies this signal and output is fed into the speaker 1. Volume control of this amplifier is also kept with the master so that that he alone can adjust the level and prevent misuse.

This method avoids some problems in the door phone application such as cross talk, feedback howling and unwieldy wiring, etc. The circuit is powered by unregulated but well filtered power supply of 9V. If it is felt that the power is a bit higher at 9V for this application, you may scale it down to 6V. 1W power available is more than adequate.

Construction

You may have to cut slots in the Veroboard to fit the fins if you are using it. Use heat sink. Use shield wire for all the interconnections from mikes, switches and volume control to amplifier. Do not attempt making speaker and mike grounds as common, as you may end up in unusual noise. Please run the wires separately and terminate at appropriate positions.

Typical DC voltages are also given in ***Figure 47,*** which will help in troubleshooting problems. These are measured under quiescent condition.

Terminal No	1	2	3	4	5	6	7	8	9	10	11	12
Voltage	8.2	4.5	8.9	0.6	0.01	GND	0.01	0.6	GND	4.5	8.2	Vcc

Figure 47

Parts

Item	*No. reqd*	*Description*	*Designation*
1	2	47uF	C1, C2
2	1	220uF	C3
3	2	100uF	C4, C5
4	2	.15uF	C6, C7
5	2	1000uF	C8, C9
6	1	2200uF	C10
7	4	4.7uF	C11, C12, C16, C17
8	2	.47uF	C13, C14
9	1	470uF	C15
10	1	Mike1	Condenser mike
11	1	Mike2	Condenser mike
12	2	BC149	Q1, Q2
13	2	82 Ohms	R1, R2
14	2	1K	R3, R4
15	2	6.8K	R5, R6
16	2	47K	R7, R8
17	2	4.7K	R9, R10
18	1	470	R11
19	2	3.9K	R12, R13
20	2	10K variable	R14, R15
21	2	4	SPK1, SPK2
22	1	IC	CD6283

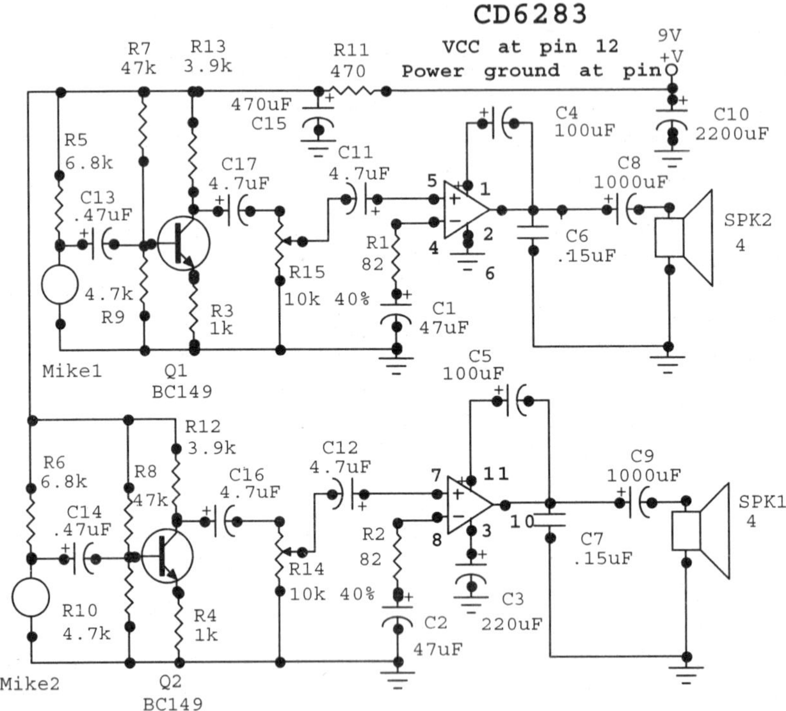

Schematic 42

Audio Power for Computer Applications

Introduction

I always had a characteristic dislike for those fancy looking speakers beside the computers. I am not convinced that these plastic speakers could ever do any justice to quality, width and depth of the audio that the computer is capable of. Then I made this amplifier and used with those good old wooden speakers. Quality? Ask me!

Description

Most of the preceding amplifiers are designed for 12V operation and cannot tackle higher voltages. Now if you are craving for little more power at a little more voltage, here is AN7147, which can handle a maximum voltage of 24Volts. It can deliver 5.5 watts in both channels into 3 ohms speakers with a gain of 44.5 dB at 12V.

IC is in the SIL package making it easy for mounting on general purpose boards. Pin out details; block diagram and PCB layout are shown here. For stereo circuit, input is given at 2 pin for left and 5 pin for right channel and output is taken out at 12 and 7 pins respectively with blocking capacitors C6 and C10.

Pin descriptions are given in **Figure 48** and block diagram is shown in **Figure 49**. Circuit is shown in **Schematic 43.**

BLOCK DIAGRAM

Pin No.	Pin Name	Pin No.	Pin Name
1	Negative Feedback Ch1	7	Output Ch2
2	Input Ch1	8	Bootstrap Ch2
3	Ripple Filter	9	Gnd (Output)
4	Gnd (Input)	10	Vcc
5	Input Ch2	11	Bootstrap Ch2
6	Negative Feedback Ch1	12	Output Ch2

Figure 48

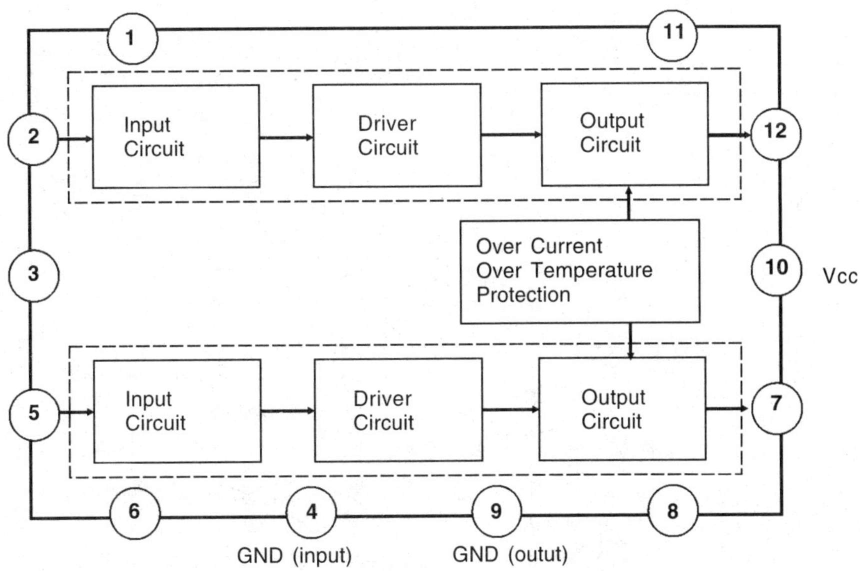

Figure 49

Construction

I used a small Veroboard for construction and the results were excellent. You need a good power supply built with adequate filtering. Minimum rating of 2 amps at 12V for the transformer is required. Transformers for the audio applications should be good, but not necessarily cheap. There should not be any air gaps in the laminations and they should not be loose. Bad transformers create uncontrollable hum in the circuit. Regulated power supply is not required in view of the large supply voltage range. Use good quality filter capacitor of about 4700 mfd. Do not compromise on these capacitors in audio applications. Bolt an aluminum heat sink to the IC.

Parts

Item	No. reqd	Description	Designation
1	3	.1uF	C1, C4, C11
2	2	1000uF	C2
3	2	47uF	C3, C5
4	2	470uF	C6, C10
5	2	100uF	C7, C8, C9
6	2	1R	R1, R2
7	2	3 Ohms	SPK1, SPK2

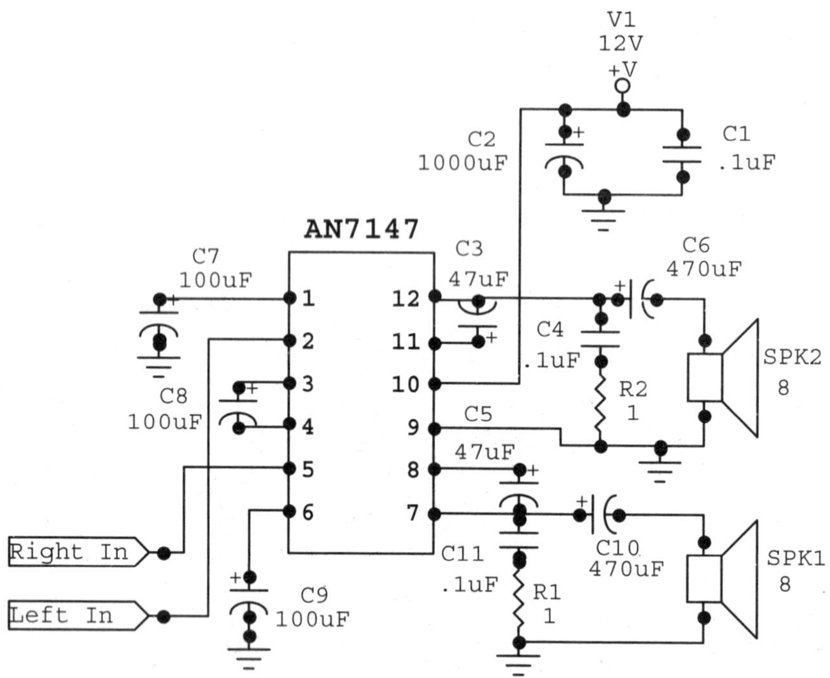

Schematic 43

Simple Radio with CA810

(Retrieve old Parts and Reduce E waste)

Introduction

It is more than 20 years after this IC hit the market but is still going strong. For a few years almost all the audio devices contained CA810, of course boasting their own figures of PMPO. It has an excellent voltage range of 3 to 18 volts. Output can reach 7 Watts at 16 volts with a suitable heat sink directly soldered on to the fins of the IC. It boasts of very high input impedance of 5M and has a gain of 51 dB. Here we have a simple AM radio built out of salvaged parts from an old pocket radio or portable transistor radio. (Well, of course, nobody stops you buying those parts from the market!) Using old parts is deliberate. There is a lot of discussion is going around the world about the menace of eWaste. Please do your might to reduce pollution.

Description

Circuit is given in **Schematic 44.** Basically a mono amplifier, this IC delivers 1W at 6V. Input is at Pin 10 through C8 and output is taken at Pin 16 through a capacitor (C7) of 1000 uF to a speaker of 8 ohms. We wish to make a simple direct conversion radio with a few additional components. Old pocket radio or a transistor must be lying around unused?! In these days of multiplying electronic garbage, I thought it would be wonderful if only a few parts could be retrieved including the IC. CA 810 IC was selected only with the view even though it is very old. The circuit can be built with any other IC easily.

Construction

First retrieval. Open an old transistor radio or a pocket radio (Wording is rather funny as all pocket radios are transistorized.) Recover the gang condenser, medium wave coil and signal diode (IN34, OA70, OA71 etc). Remove the gang condenser. It is the variable condenser which is used to tune various stations with a knob or pulley fixed to it. Desolder all the three terminals, open the screws and pull it out with the knob. Next is the medium wave coil. It is easy to identify this as this has more number of turns of thin enamel or litz wire coiled around ferrite rod. Carefully remove all three or four wires without snapping them. Remove it along with ferrite rod. Remove also signal diode in the detector stage or from AGC. This is generally a small little glass diode in the vicinity of last IFT or before volume control. If you wish to buy all these components, fine. See the parts list.

Construction of 810 Amplifier is simple and straight forward. You may follow the general instructions given in the earlier circuit for populating PCB and mounting heat sink. There is a little difference in the pin-out (the way in which the pins are numbered) of CA 810 and its Japanese cousin TBA 810. Follow the pin out of whichever IC you are using.

As it can be observed from the schematic, front end does not require power at all. It pulls local medium wave easily. We now have too many of them around. Carefully slide the MW coil on the ferrite rod for best reception. If necessary add an additional wire as an antenna.

You can also connect FM radio to the volume control. Full fledged FM or AM radios are not discussed here as it would be difficult for a hobbyist to make those antenna, oscillator and IF coils. However you may buy a ready made FM module and connect its output to the volume control.

Parts

Item	No. reqd	Description	Designation
1	3	100uF	C1, C2, C3
2	1	1kpf	C4
3	1	33kpf	C5
4	5	.1uF	C6, C8, C10, C12, C13
5	2	1000uF	C7, C9
6	1	100	R1
7	1	56	R2
8	1	1	R3
9	1	100K variable	R4
10	1	4 Ohms	SPK1
11	1	CA810	U1

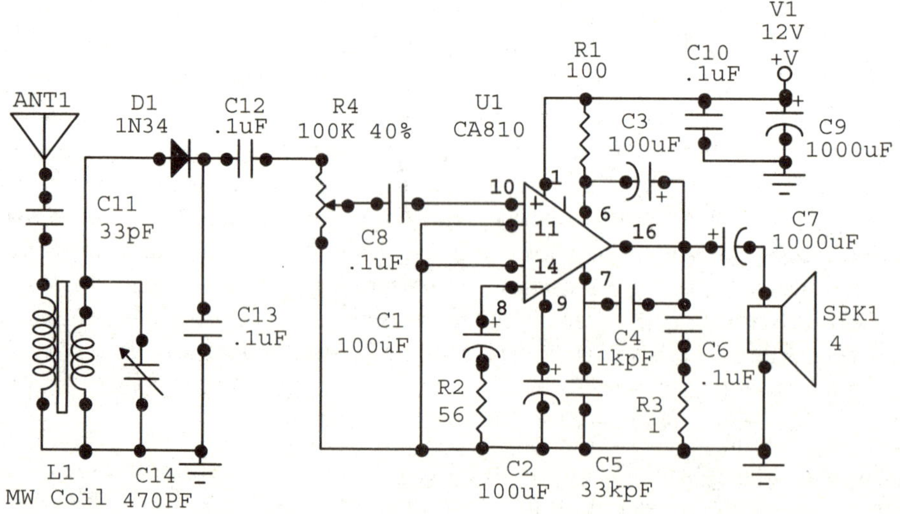

Schematic 44

10W Audio Amplifier

Introduction

Another audio device from Phillips primarily designed for car applications which can deliver up to 10W. It can stand voltages up to 24 and hence it is highly suitable for mains application.

Description

Circuit is given in **Schematic 45.** IC is housed in 9-pin single line-package, which makes it convenient for mounting on the veroboards and fixing heat sinks. As TDA 1010 is only a mono amplifier, you have to use two for stereo applications. It can deliver 6.4 watts with 2 ohms load, or 6.2 watts with 4-ohm load at 14.4 volts supply. It can also deliver 9 watts with additional bootstrap resistance between Pins 3 and 4 at 14.4 volts supply with 2 ohms load. With a voltage gain of 54, it has excellent ripple rejection; preamplifier and main amplifier are internally separated, allowing certain flexibility.

Construction

I used a small Veroboard for construction and the results were excellent. You need a good power supply built with adequate filtering. Minimum rating of 1 amp at 12V for the transformer is required for single channel application. Transformers for the audio applications should be good, but not necessarily cheap. There should not be any air gaps in the laminations and they should not be loose. Bad transformers create uncontrollable hum in the circuit. Regulated power supply is not required in view of the large supply voltage range. Use a good quality filter capacitor of about 4700 mfd. Do not compromise on these in audio applications. Bolt an aluminum heat sink to the IC.

Parts

Item	*No. reqd*	*Description*	*Designation*
1	4	.1uF	C1, C3, C5, C9
2	1	100uF	C2
3	1	.47uF	C4
4	1	1kpf	C6
5	2	1000uF	C7, C8
6	1	330K	R1
7	1	1	R2
8	1	1M variable	R3
9	1	4 Ohms	SPK1
10	1	TDA1010/1011	U1

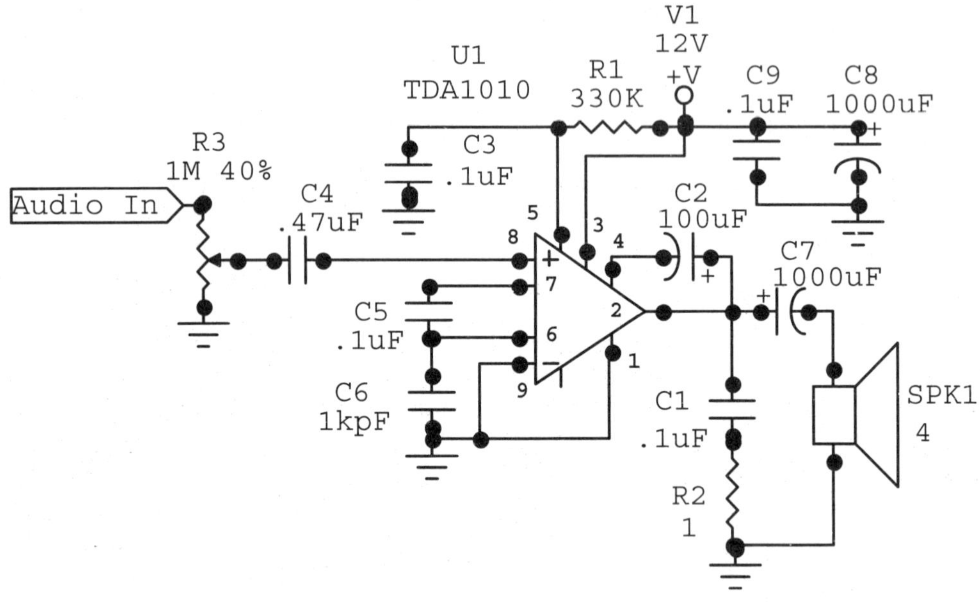

Schematic 45

20W Amplifier

Introduction

If you are craving for more power, here is another popular and wonderful device from Sanyo capable of pure RMS 6W in stereo configuration and pure RMS 19W in bridge (typical), LA4440 can stand voltages up to 24 and hence it is highly suitable for mains application.

Description

LA 4440 has excellent features with minimum number of external parts required and good ripple rejection 46dB (typ.) and channel separation. IC has low distortion (0.1%) over a wide range from low frequencies to high frequencies. Heat sink radiator is easy to mount at the back of IC with two screws. It boasts of protectors against over voltage at a set point of 25V. It is capable of withstanding up to 50V at giant pulse surge for 200milliseconds. IC is protected even when power is applied in a state with pins 10, 11, and 12 short-circuited with solder bridge, etc. Since the input circuit uses PNP transistors and the input potential is designed to be zero bias, no input coupling capacitor is required and direct coupling is available.

Above all it has built-in 2 channels (dual) amplifiers for use in stereo capable of 6W per channel and bridge amplifier applications at 19 W RMS. Small pop noise at the time of power supply ON/OFF should not matter much.

Configuration of the bridge amplifier is shown below, in which ch1 and ch2 operate as non-inverting amplifier and inverting amplifier respectively. The output of the non-inverting amplifier divided by resistors R3, R4 is applied, as input, to the inverting amplifier. Since attenuation (R4/R3) of the non-inverting amplifier output and amplification factor (Rf/R4+RNF) of the inverting amplifier are fixed to be the same, signals of the same level and 180^0 out of phase with each other can be obtained at output pins (12) and (10). Circuit is given in **Schematic 46.** Pin voltages are shown in ***Figure 50*** and block diagram in ***Figure 52.*** Configuration of Bridge amplifier is shown in ***Figure 51.***

Let us have a word on various external components and their use, which makes the understanding and later trouble shooting of the IC easy.

Description of External Parts

C1 (C2) · Feedback capacitor: The low cutoff frequency depends on this capacitor. If the capacitance value is increased, the starting time is delayed.

C3 (C4) · Bootstrap capacitor: If this capacitance value is decreased, the output at low frequencies gets decreased.

C5 (C6) · Oscillation preventing capacitor: Polyester film capacitor, should be used for its good temperature and frequency characteristics.

C7 (C8) · Output capacitor: This capacitor dictates the low cutoff frequency. In the bridge amplifier mode, the output capacitor is also generally connected.

C9 · Decoupling capacitor: Used for the ripple filter. Since the rejection effect is saturated at a certain capacitance value, it is meaningless to increase the capacitance value more than required. This capacitor affects the starting time as it is also used for the time constant of the muting circuit.

C10 · Power source capacitor.

R1 (R2)· Filter resistor for preventing oscillation.

R6 (R7). Used at bridge amplifier mode in order to increase discharge speed and to secure transient stability.

R3 (R4)· Resistor for making input signal of inverting amplifier in Voltage Gain Adjust at Bridge Amplifier.

PIN VOLTAGES

TYPICAL DC VOLTAGES AT EACH TERMINAL Vcc = 13.2V														
Terminal No	1	2	3	4	5	6	7	8	9	10	11	12	13	14
Voltage	1.4	0.03	0	0	13	0.03	1.4	0	11.9	6.8	13.2	6.8	11.9	0

Figure 50

Configuration of the Bridge Amplifier

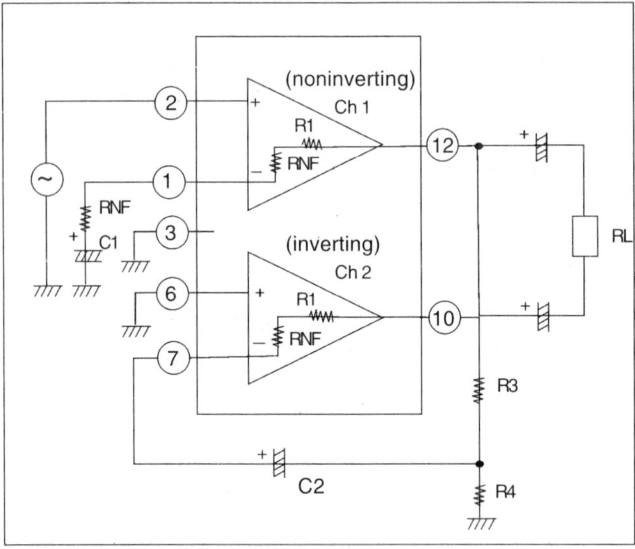

Figure 51

Block Diagram

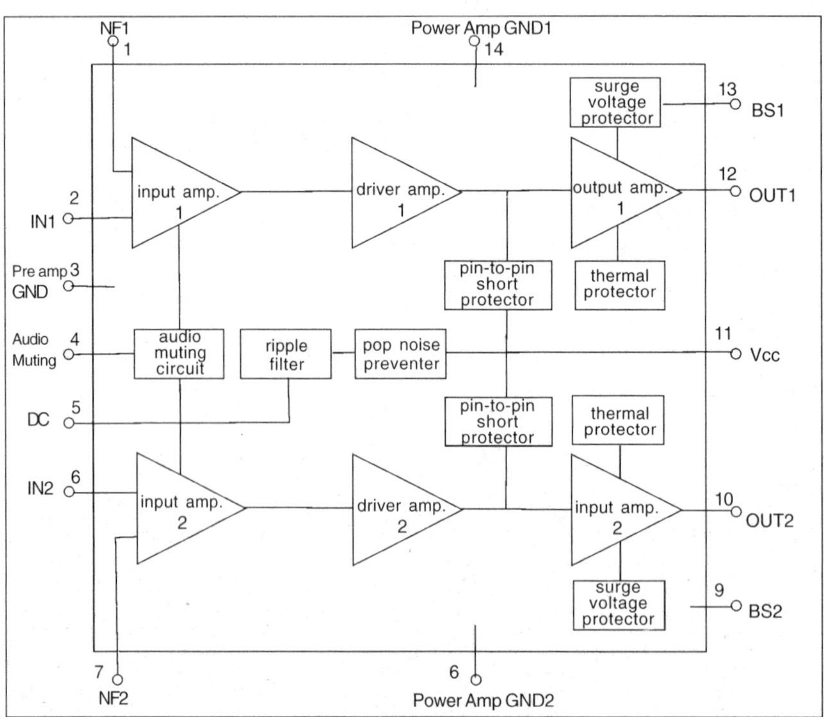

Figure 52

Construction

I used a small Veroboard for construction and the results were excellent. You need a good power supply built with adequate filtering. Minimum rating of 2 amp at 12V for the transformer is required for single channel application. Use a good quality filter capacitor of about 4700 mfd. Do not compromise on these in audio applications. Bolt an aluminum heat sink to the IC and fasten the heat sink firmly to a chassis or the enclosure. Solder pins of the IC only after heat sink is mounted. Pin voltages in quiescent mode are given. This will help in trouble shooting. Take the readings when there is no signal.

Parts

Item	No. reqd	Description	Designation
1	2	47uF	C1, C2
2	2	100uF	C3, C4
3	2	.1uF	C5, C6
4	3	1000uF	C7, C8, C10
5	1	220uF	C9
6	2	1	R1, R2
7	1	16K	R3
8	1	160	R4
9	2	430	R6, R7
10	1	4	SPK1

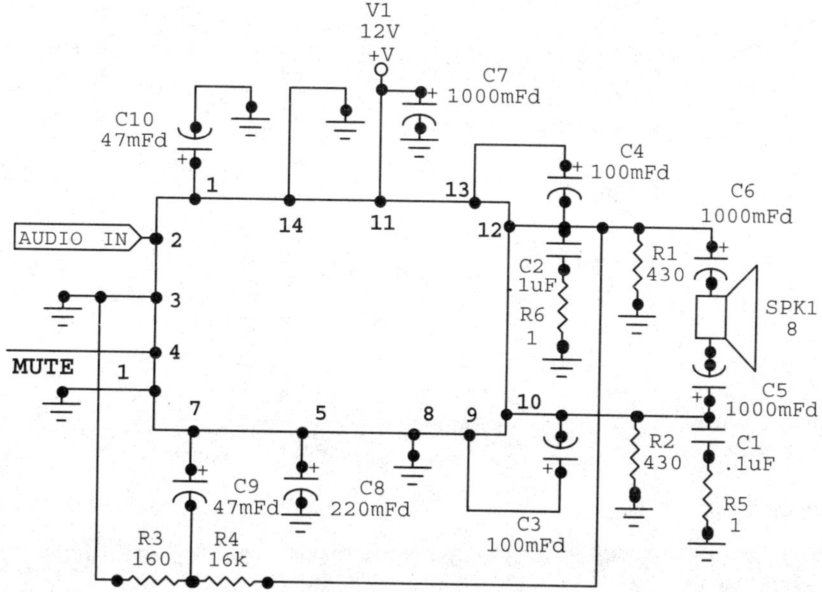

Schematic 46

Audio Mixer

Introduction

Here we have an excellent audio mixer which has four mike inputs and two auxiliary inputs. This is useful when a number of inputs are to be connected to a power amplifier or as an additional source of inputs. We have an audio out for feeding to the main amplifier and a line out for other purposes.

Description

An op-amp is basically a differential amplifier with a large voltage gain, very high input impedance and low output impedance. The op-amp has a "inverting" or (-) input where the phase of the input signal is inverted and "non -inverting" or (+) input where the phase of the input signal is not inverted when it comes at the output. Pin out of LM324 is given in Figure 53.

PIN OUT

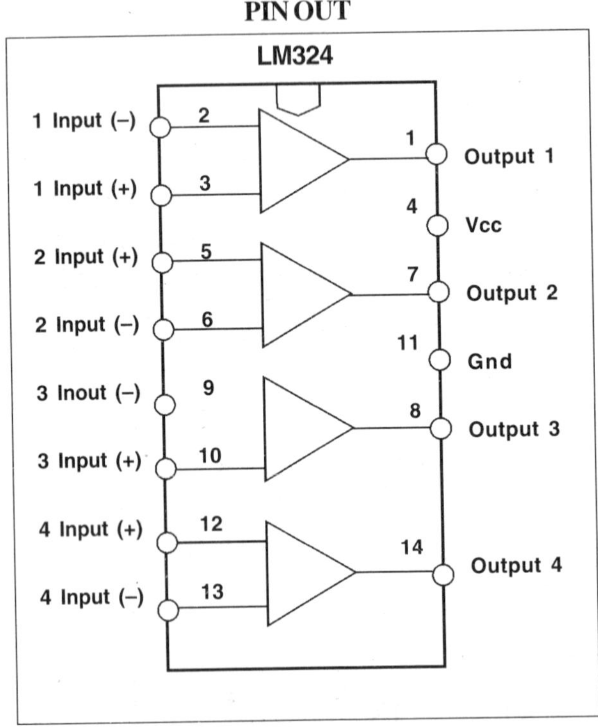

Figure 53

Cirduit is shown in Schematic 47. LM324 has four independent operational amplifiers in a DIL package which can be operated from a single supply voltage in the range of 3 to 32V. Each of them is configured as an inverting amplifier with a gain of 10 and input impedance of 10K.

The output levels of these amplifiers are controlled by 10K variable controls. Then the outputs are directed to two places. One goes to summing amplifier with a gain of 4. Another goes to a display device, detailed description of which is given in the next project. There are also two line inputs. As these have enough signal strength, no additional amplification is necessary and they go directly to volume controls. Next IC is dual op amp MC1458. Pin out of MC1458 is given below.

PIN OUT (Top View)

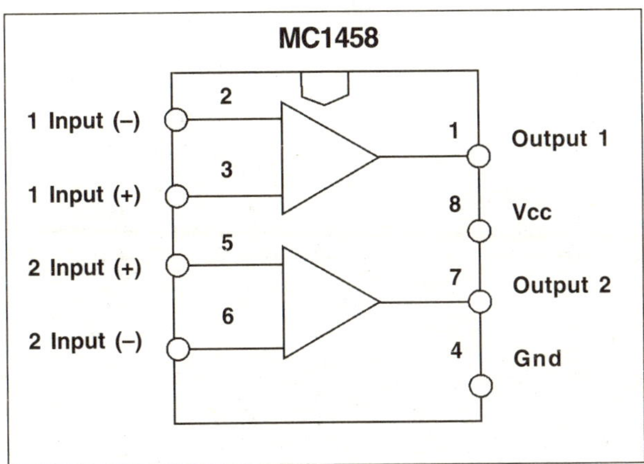

Figure 54

MC1458 is high performance monolithic dual operational amplifier intended for a wide range of analog applications, like summing amplifier, voltage follower integrator, active filter and function generator. The high gain and wide range of operating voltages provide superior performance in integrator, summing amplifiers and general feedback applications.

All the controlled levels from six volume controls are fed into one summing amplifier of MC1458. Output of this IC is taken as the AUDIO OUT to be fed into the main amplifier. Another Op amp in this IC also acts as another summing amplifier with a gain of 2 and its output is taken as LINE OUT. There is an LED at its output which gives a general indication of the mixer performance. There is another feature which can be implemented in conjunction with the next circuit to display the actual level of each mike. There are four terminals marked OUT. These can be inputted to two ICs of TA 7666 stereo LED display IC as shown in the next. This can display levels of each mike. Please ignore it if not required.

Construction

You will need RCA sockets for all input and outputs. The Mixer should be enclosed in a metal box. The power supply should be well filtered and preferably regulated. Use shield wire for all the input, output and volume control connections. Use of sliding volume controls is helpful, as they give direct indication of the operating level.

Parts

Item	No. reqd	Description	Designation
1	10	10uF	C1, C2, C3, C4, C5, C6, C7, C8, C9, C10
2	1	LED1	D1
3	4	100K	R1, R2, R3, R4
4	9	10K	R5, R6, R7, R8, R10, R14, R16, R18, R19,
5	4	10K variable	R9, R12, R13, R17, R22, R23
6	1	47K	R15
7	1	22K	R20
8	1	1K	R21
9	4	LM324	U1
10	2	MC1458	U2

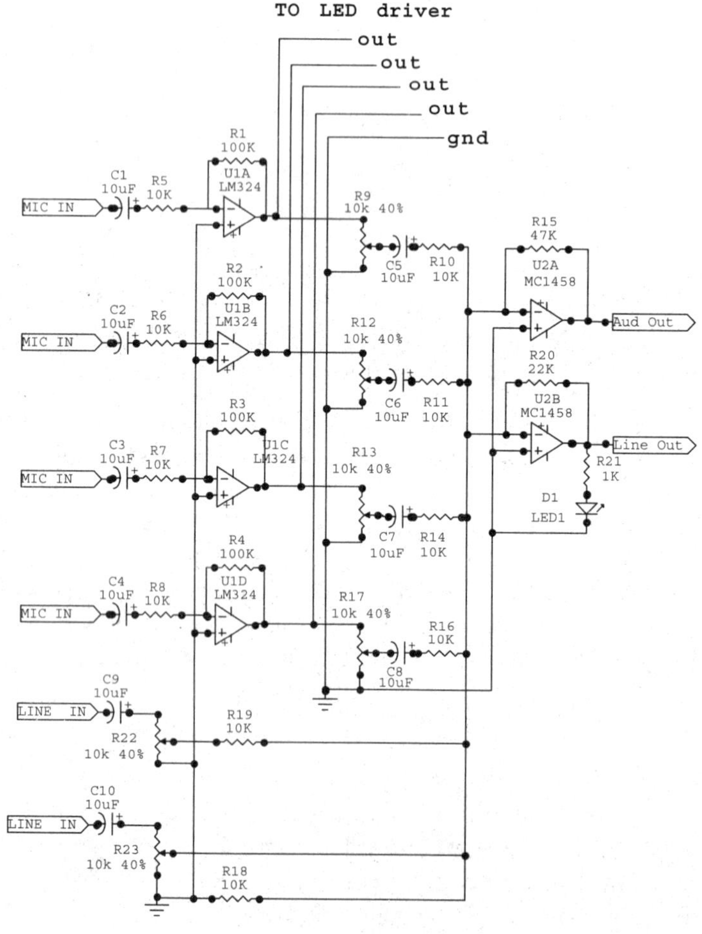

Schematic 47

Mike Mixer Level Display / Stereo Display

Introduction

The following circuit is appended to the previous circuit but can also be used individually to display audio levels. TA7666 is a stereo display driver displaying audio levels in 5 LEDs each. Two of these ICs will be required to display the mixing levels of all the four mike inputs and six of them are required for all the inputs in the earlier circuit. But a single IC can display stereo output levels and it can be used as such.

Description

Circuit is shown in Schematic 48. TA7666 has two inverting amplifiers, ten comparators, and a reference voltage network for stereo display applications. It has a wide supply range of 6 to 12 volts and quiescent current of 4mA. Because of the internal inverting amplifier voltage gain can be adjusted. Only one IC is shown for the purpose of simplicity but two such ICs are required to display four mike channels and three will be required if one wishes to display even the line in channels.

Construction

Take the four wires coming out of earlier mike mixer OUTPUTS and connect them to the inputs of both these ICs. Connect ground to ground. Use good quality shield wire for interconnection. Fine if you are using RCA sockets and pins. If you do not wish to use it along with the mike mixer, just ignore it and make a straight stereo display.

Parts

Item	*No. reqd*	*Description*	*Designation*
1	4	4.7uF	C1, C2,
2	2	0.47uF	C3, C4
3	10	LED1	D1, D2, D3, D4, D5, D6, D7, D8, D9, D10
4	10	1K	R1, R3, R4, R5, R6, R7, R8, R9, R10,
5	1	470	R2
6	2	39K	R12, R13
7	2	10K	R14, R15
8	1	U1	TA7666

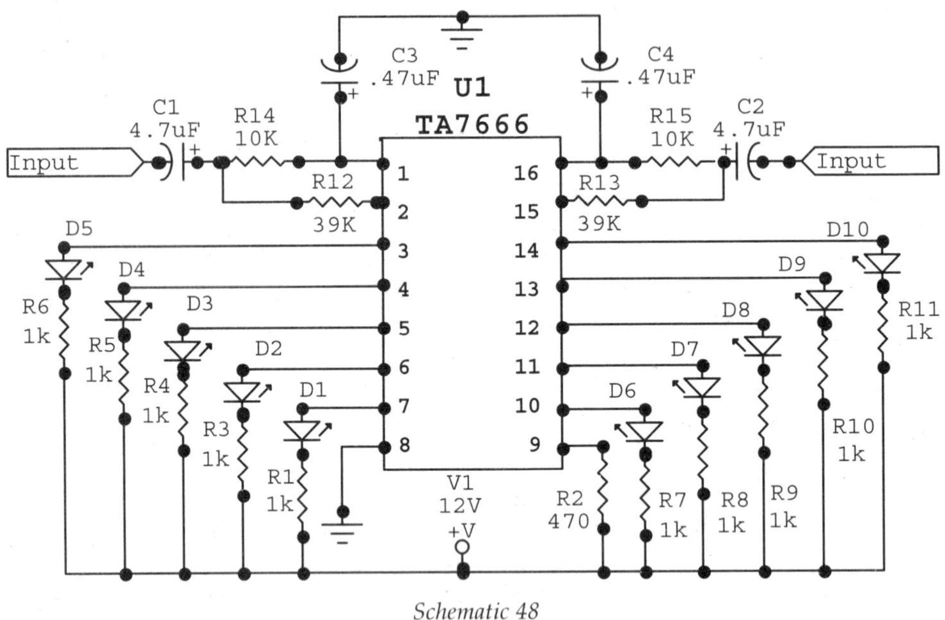

Schematic 48

LB1405 Display

Introduction

LB 1405 is an individual audio display to visualize audio levels in 5 discrete LEDS. You will need two such ICs to make it useful for stereo application. This can be used also as DC level meter such as signal meter or voltage monitor.

Description

Circuit is shown in Schematic 49. LB1405 features a good supply range of 4.4V to12V. Due to built-in constant voltage circuit, output levels are maintained constant in spite of voltage fluctuations. IC has high input impedance and response time can be adjusted externally. The following application is for display of levels at speaker output. If you wish to use it for stereo, use two ICs. These ICs of course can be cascaded to nine LEDs on display.

Construction

Construction of this project should not pose any problem. Use of Vero board is OK. Two wires from the speakers are taken and inputted at Pin 3 and ground appropriately. Adjust level control with R2.

Parts

Item	No. reqd	Description	Designation
1	1	120pF	C1
2	2	4.7uF	C2, C4
3	1	1uF	C3
4	5	LED1	D1, D2, D3, D4, D5
5	1	100K	R1
6	1	22K Preset	R2
7	1	24K	R3
8	1	1.5K	R4
9	1	33K	R5
10	1	18K	R6
11	1	U1	LB1405

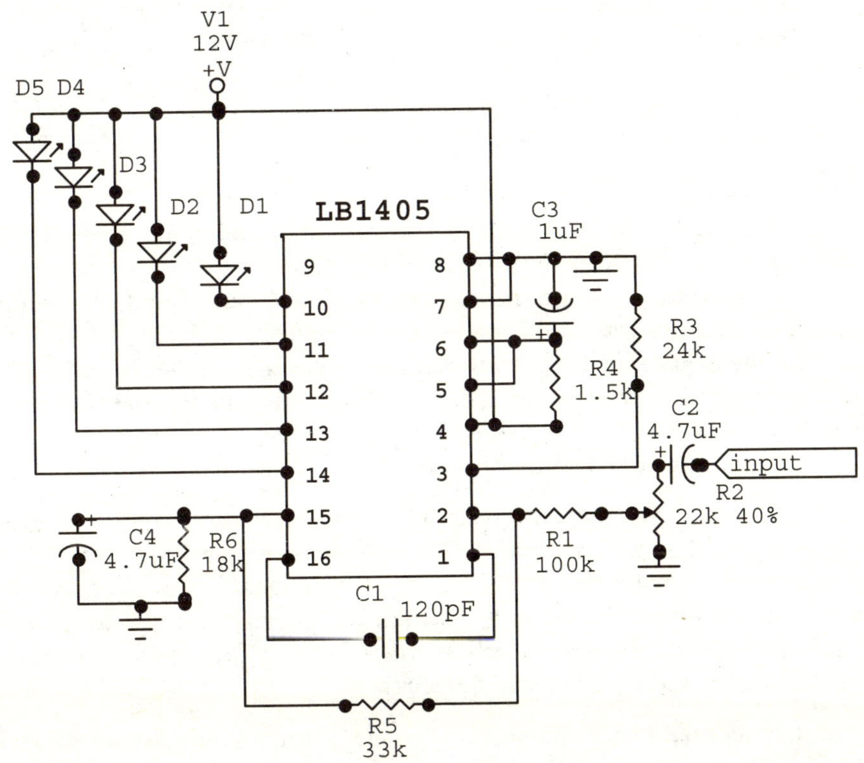

Schematic 49

Recording Voice On IC

Introduction

We have come a long way from the Edison's "Mary had a little lamb!" These are the days of recording directly onto the chips. No spools, cassettes, CDs or no running motors! Voice chips are slowly invading us in various daily chores, elevators, microwave ovens, washing machines; answering machines and are entering forbidden places like mantras, slokas and temples. Most of them are one time recordable devices requiring special software to record the original messages. But here we have a re-recordable and individual voice IC in APR9600. It is a high-quality voice recorder / playback IC with capability for 32 to 60 seconds. It is easy to build and all the parts are generally available.

This IC can record and playback a number of messages at random in 2, 4, or 8 segments, or serially in a tape mode with the messages recorded and played back one after the other just like in cassette tape. There are various sample rates provided for the IC. If you select high sample rates, you can get quality recording but with a less total time. Even with normal sample rates you can still get excellent recording up to one minute. Microphone amplifier, AGC circuits, internal anti-aliasing filter, integrated output amplifier and message management are all built into IC.

Description

APR9600 has a 28 pin DIP package with a supply voltage between 4.5V to 6.5V. During recording and replay, current consumption is 25 mA and in idle mode, it drops to 1 uA. In this low-cost high performance sound record/replay IC incorporating flash analogue storage technique, recorded sound is retained even after power supply is removed from the module. IC can operate in one of two modes: serial mode and parallel mode. In serial access mode, or tape mode sound can be recorded in 256 sections. In parallel access mode, sound can be recorded in 2, 4 or 8 sections. IC is operated by simple push button switches. It can also be controlled by micro-controllers and computers.

Total sound recording time can be varied from 32 seconds to 60 seconds by changing the value of a single resistor. At a sampling rate of 4.2 kHz recording period is 60 second with a bandwidth of 20Hz to 2.1 kHz. However, higher sampling rate of 8.0 kHz can be achieved but it reduces recording time to 32 seconds. Evidently higher sampling rates give better voice quality but they also increase the bandwidth. 12 mW of power is available from the chip which can be directly connected to a speaker of 16 ohms.

The Pin out and functional block diagram of the device are given ***Figure 55*** and ***Figure 56*** respectively. Chip is managed by the device control block (top right side) as seen in the functional diagram and messages are managed through the message control block (bottom right).

PIN OUT

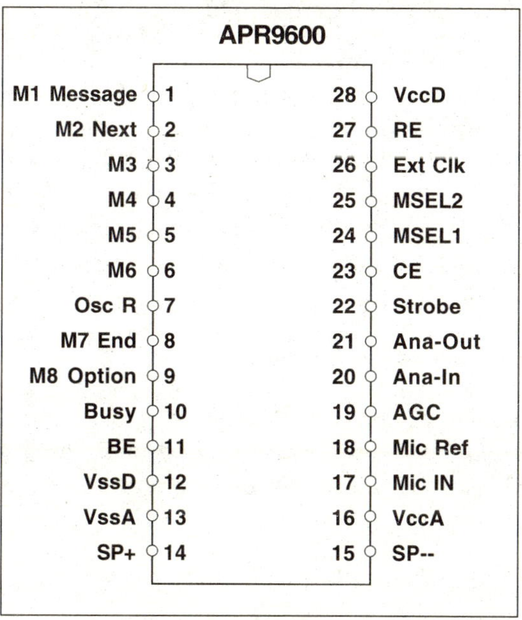

Figure 55

BLOCK DIAGRAM

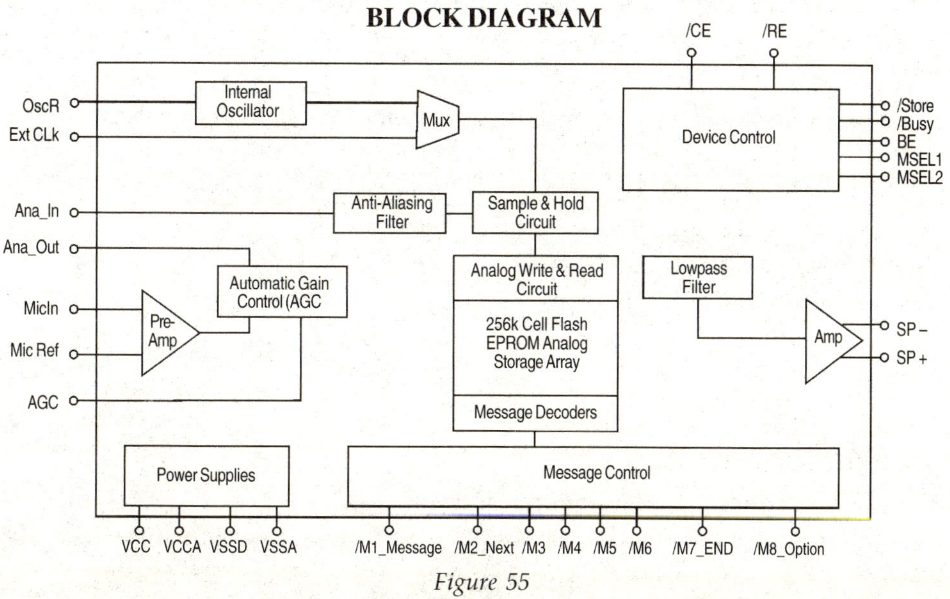

Figure 55

Recording

Voice signal from the Electret microphone is fed into the device and amplified by a preamplifier. Then the signal is processed by an automatic gain control circuit. It is further amplified by connecting the Ana_Out pin (21) to the Ana_In pin (20) through an external capacitor. Both pins Mic.In (18) and Mic.Ref (19) must be coupled to the microphone through capacitors.

Then the signal is passed through an internal anti-aliasing filter which automatically adjusts its response according to the sampling frequency selected. The signal is now routed into the memory array through a combination of the Sample and Hold circuit and the Analog Write/ Read circuit. APR9600 samples incoming voice signals and stores the instantaneous voltage samples in non-volatile FLASH memory cells in 8-bit (256) binary encoded values. Recording from external source can also be done directly into the Ana_In pin (20) through a capacitor, but Ana_In (20) and Ana_Out (21) pins should still be coupled for playback.

Playback

During playback the stored signals are fetched from memory, smoothed to form a continuous signal. After a low pass filter, they are amplified. Then the signals are fed into the speaker terminals, SP+ (14) and SP- (15) pins and can be heard at about 12mW power at 16 Ohms.

Management

Clock frequency is provided for by an internal oscillator, which can be changed by changing the resistance at the Osc R pin(7) to GND. Table (***Figure 57)*** prescribes the resistance values and the consequent sampling frequencies, and the resultant bandwidth and duration of recording. RC network on pin 19 sets the AGC "attack time".

Ref Rosc	Sampling frequency	Input Bandwidth	Duration Seconds
84K	4.2 kHz	2.1 kHz	60
38K	6.4 kHz	3.2 kHz	40
24K	8.0 kHz	4.0 kHz	32

Figure 57

Message Management

The chip can be recorded in the following modes

- **Parallel or Random access mode with 2, 4, or 8 messages within the total recording time.**
- **Tape mode, with two options: - Auto rewind and Normal.**

These are defined by Pins MSEL1 (24), MSEL2 (25), /M8 OPTION (9) as shown in table below. Modes cannot be mixed, and should not be changed while recording. APR9600 prompts with audible beeps at the output about the changes in the device's status, when BE pin (11) is logic high. Modes are given out in ***Figure 58.***

Mode	MSEL1	MSEL2	/M8 OPTION
Random Access 2 fixed duration messages	0	1	Pull this pin to Vcc through 100K resistor
Random Access 4 fixed duration messages	1	0	Pull this pin to Vcc through 100K resistor
Random Access 8 fixed duration messages	1	1	Becomes the /M8 message trigger input pin
Tape mode Normal Operation	0	0	0
Tape mode Auto rewind Operation	0	0	1

Figure 58

General Functional Description

On power up, /CE (23) must be pulled low to enable the device (S1). RE (27) is low for recording and high for playback (S2). As the recording begins IC prompts with a beep. Recording/playback starts by using the switch of the corresponding trigger pin as per the following description. Recording stops automatically with two beeps, if the trigger pin is held low beyond the end of the maximum allocated duration. Busy (10), Strobe (22), or M7_END (8) pins have LEDs as indicators of device status. Busy LED indicates that the device is busy and that no commands can be currently accepted. Strobe LED pulses when a memory segments is used. The APR9600 has a total of eighty memory segments. M7 END LED is the stop indicator.

Construction

Assembly is straightforward and I made it on a piece of 2" X 4" Vero board. It will be easy to use an IC base for the main chip. Here the circuit diagrams for all the three modes are given. Out put from pin 14 can be driven into any of the audio amplifiers discussed earlier after by passing Pin 15 as shown. Amplifier drive is shown in Random access mode. A small speaker of 8 to 16 ohms also can be driven directly as shown in tape mode. Speakers should be farther away from the mike to avoid annoying howling.

These chips can be had from Aplus India, Mumbai.

1. Random Access mode

Recording can be made in 2, 4, or 8 sections as per selection made on MSEL1, MSEL2 and M8 pins. **Schematic** 50 shows recording in eight sections.

Recording in two sections

Pull MSEL1 (24) to low level and MSEL 2 (25) to high level and option pin 8 is pulled up by 100k resistance. Each segment records by half of total recording time.

Recording in four sections

Pull MSEL1 to high level and MSEL 2 to low level and option pin 8 is still pulled up by 100k resistance. Each segment records a quarter of the total recording time.

Recording in eight sections

Pull both MSEL1 and MSEL 2 to high level and option pin 8 becomes the rigger pin. Each segment records a eighth of the total recording time.

Schematic shows record/playback in eight sections. For two and four sections unused message trigger pins are left unconnected. But pin M8_Option (8) should be pulled to VCC through a 100k resistor.

The message segment trigger pins are marked M1_Message - M8_Option from pins 1 to 9 (excluding pin 7) for message segments 1-8 respectively. To record pull the recording / playback switch at RE (27) low with switch S2, and using any of the switches S3 to S10 corresponding to the intended message trigger pin, pull it low. Start recording by speaking into the microphone. As the recording begins, IC prompts with a beep. Recording stops automatically with two beeps, if the trigger pin is held low beyond the end of the maximum allocated duration.

Playback

To playback, recording /playback switch at RE (27) high with switch S2. Press the switch S3 to S10 corresponding to the required message trigger pin pulling it low. Playback will continue until the end of the message is reached. Playback will stop if the same switch in S3 to S10 is pressed again. If a different switch is pressed during playback, playback of the present message stops immediately (indicated by one beep) and playback of the new message begins.

Parts

Item	*No. reqd*	*Description*	*Designation*
1	1	1uF	C1
2	1	4.7uF	C2
3	4	.1uF	C3, C4, C6, C12
4	2	22uF	C5, C7
5	1	100uF	C13
6	2	LED	D2, D3
7	1	38K	R1
8	2	680	R2, R4
9	1	100K	R5
10	1	220K	R6
11	2	4.7K	R8, R9
12	1	1K	R10
13	9	Push to switches	S1, S3, S4, S5, S6, S7, S8, S9, S10,
14	1	SPST	S2
15	1	ELECTRET MIKE	X1

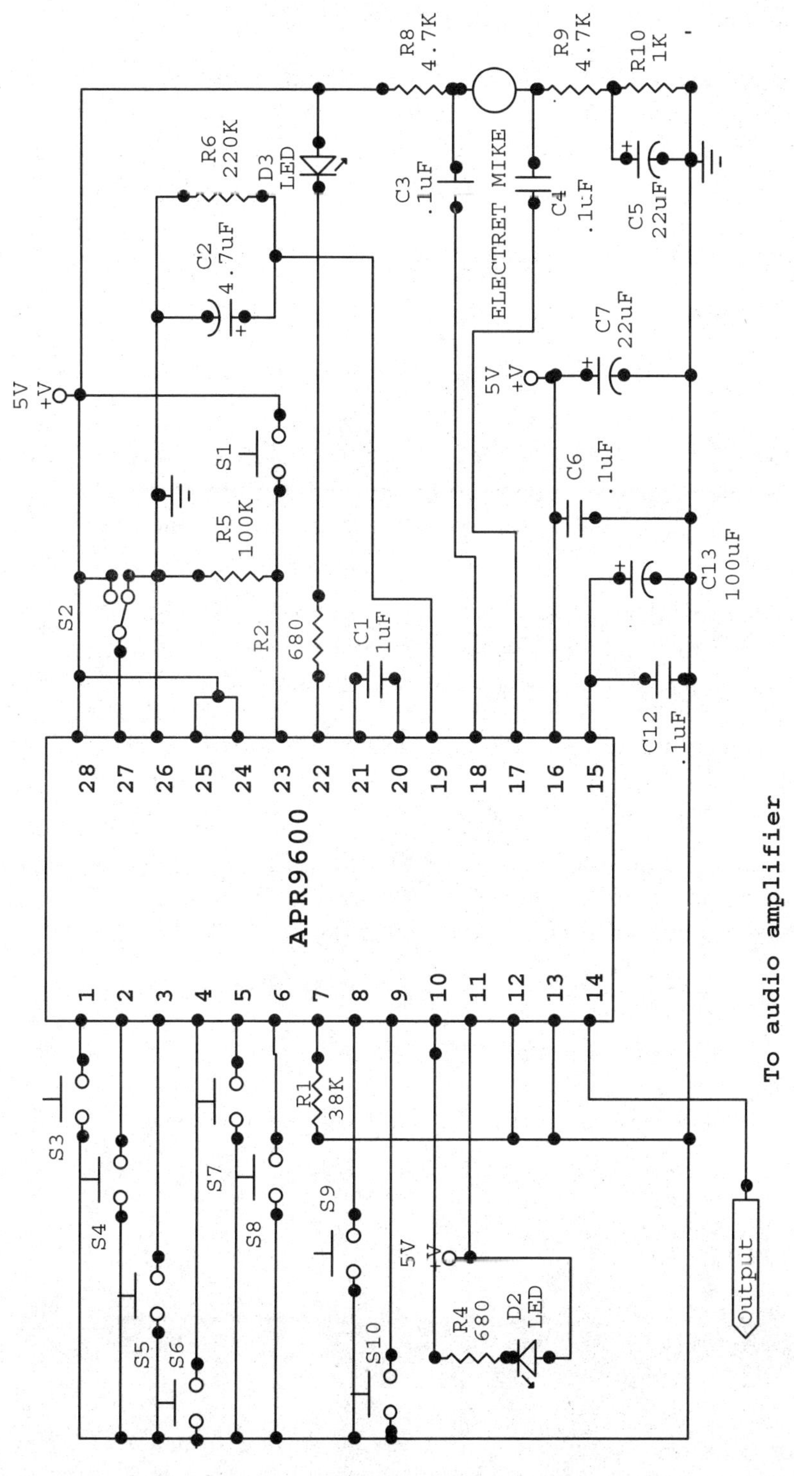

Schematic 50

2. Tape Mode

Tape mode or serial mode is like cassette tape recorders, but has two options, auto rewind and normal. Obviously in auto rewind mode IC automatically rewinds to the beginning of the message immediately after recording or playback of the message. In normal mode it must be switched for a rewind.

Tape Mode using the Auto Rewind Option

Recording

Pull both MSEL1 and MSEL 2 to low level and option pin 8 to high level. RE (27) must be low for recording (S2).

Press /M1_Message segment pin (1) low with switch S3 to start voice recording. Hold it and speak into the mike as long as required but within the time limits set. Release S3 to stop recording and the device will automatically rewind to the beginning of the most recent recorded message.

Now if you press S3 (M1) is pressed again and released, the sound track counter will not go forward to the next sound track location. To fast forward S4 (M2) should be toggled. Hence to record further, pull /M2_Next pin (2) low with S4. If this is not done, a subsequent recording will overwrite the last recorded message. Each sound track may have different lengths, but the accumulated length of all sound tracks will not exceed total time say, 60 seconds

To record over a specific message, pulse (S1) CE (23) pin low once and rewind to the beginning of the voice memory. Using the switch S4, M2_Next (2) pin for a required number of times and reach the beginning of the message to be erased. Then start recording as above but all previous messages beyond this will be erased now.

Playback

RE (27) must be HIGH to play back (S2) and press and release S3 (–M1) to listen to the recorded message. Press S4 again and again to play the 2nd, 3rd, 4th and other consecutive sound tracks. Press S1 (CE) to reset the sound track counter to zero. This mode is shown in **Schematic 51.**

Construction

Assembly is straightforward and I made it on a piece of 2" X 4" Vero board. It will be easy to use an IC base for the main chip.

Parts

Item	*No. reqd*	*Description*	*Designation*
1	1	1uF	C1
2	1	4.7uF	C2
3	4	.1uF	C3, C4, C6, C12
4	2	22uF	C5, C7
5	1	100uF	C13
6	3	LED	D1, D2, D3
7	1	38K	R1
8	3	680	R2, R3, R4
9	1	100K	R5
10	1	220K	R6
11	2	4.7K	R8, R9
12	1	1K	R10
13	3	Push to On switch	S1, S3, S4
14	1	SPST	S2
15	1	ELECTRET MIKE	X1

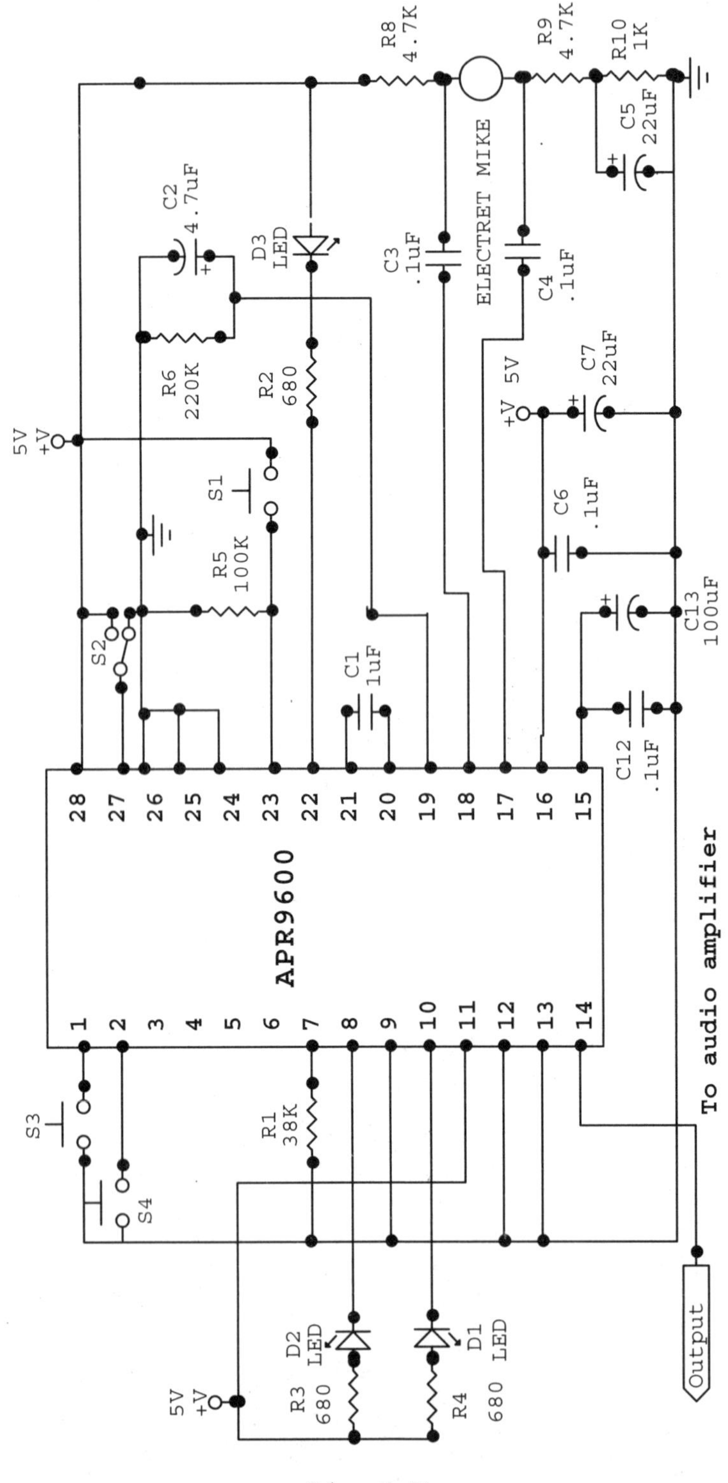

Schematic 51

3. Tape Mode Normal

Recording

Pull MSEL1, MSEL 2 and option pin 8 to low level. RE (27) must be low for recording (S2). Press S3 (1) corresponding to M1_Message segment pin low to start voice recording and hold it as long as required. If the M1_Message pin is held low beyond the end of the available memory, recording stops automatically with two beeps. Pressing S3 (1) again starts a new recording after the last recorded message, thus preserving it. To record over all previous messages, press S1 to make CE pin (23) low once and reset the device. The most recently recorded message will become the last recorded message and all previously recorded messages following this message will be erased. If you wish to preserve any recording, use Auto Rewind option instead of Normal option.

Playback

Play back mode is much the same except that the RE must be HIGH.

This mode is shown in **Schematic 52.**

Construction

Assembly is same as earlier modes.

Parts

Item	*No. reqd*	*Description*	*Designation*
1	1	1uF	C1
2	1	4.7uF	C2
3	4	.1uF	C3, C4, C6,
4	2	22uF	C5, C7
5	3	LED	D1, D2, D3
6	1	38K	R1
7	3	680	R2, R3, R4
8	1	100K	R5
9	1	220K	R6
10	2	4.7K	R8, R9
11	1	1K	R10
12	2	Push to On switch	S1, S3
13	1	SPST	S2
14	1	ELECTRET MIKE	X6

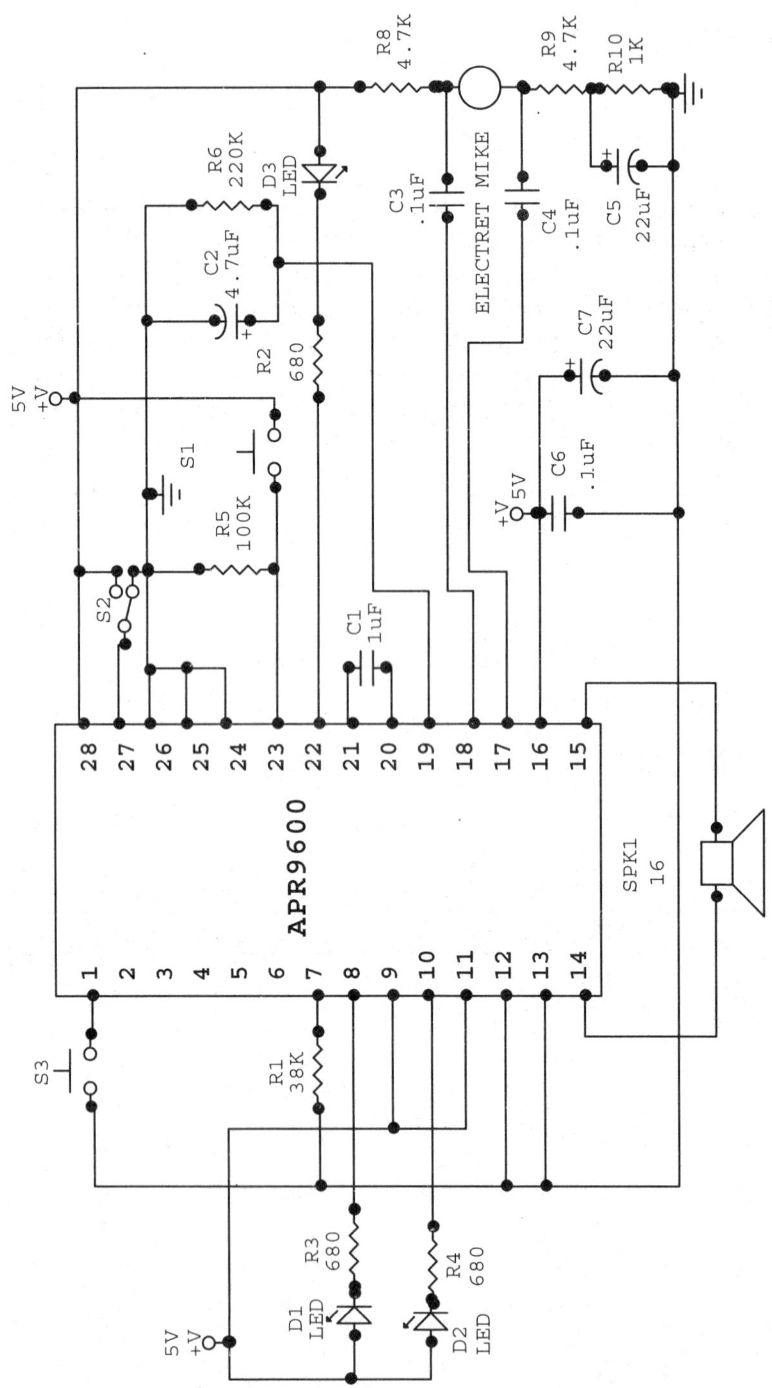

Schematic 52

Single Digit Counter / Live Mains Indicator

We are beginning a series of counters starting from single digit display to four digits. First two makes use of CMOS counter IC CD 4033. This IC has a decade counter, seven segment display decoder and driver all in one package. It has ripple blanking and lamp test features with a supply range of 3-18V. While there are a number of display drivers available in various logics and combinations, this was selected because of its simplicity, availability and direct interface with displays.

Introduction

We now have a wireless mains indicator using this IC. This can also be used as a single digit counter. With the following project we would be able to detect buried or concealed mains wiring and also detect breakages in the wiring. Presence of mains is shown as rapidly changing display from zero to nine at the frequency of mains (50 Hz). Probe is made of simple coil of wire which is brought near the mains wiring.

Do not; do not ever connect this circuit directly to mains.

Description

CD4033 is a direct drive IC for seven segment LED displays with all the necessary features built into it. It has features to cascade more number of ICs to drive more number of displays. IC has clock (1), clock inhibit (2), and reset (15) inputs and seven decoded (10, 12, 13, 9, 11, 6, 7) and carry out (5) outputs. CD4033 also has ripple blanking input (3) and output (4) and a lamp test (14) facility. Counter is advanced by positive transition of a signal at the clock input when the clock inhibit is low. Decoded outputs go high as per the clock input making the corresponding segments glow. A high level at the reset pin clears the counter to zero.

Carry out signal completes one cycle for every ten clock cycles and is used to cascade ICs to increase the number of digits. If the lamp test pin is made high, it enables all the outputs. This will make all the segments to glow and can be used to check up of seven segment display. Ripple blanking input and output help in blanking out non-significant digits.

Pin out of the IC is given in **Figure 59.**

PIN OUT

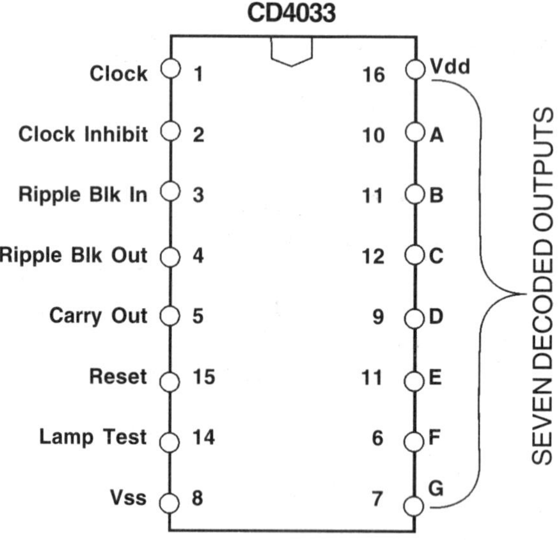

Figure 59

We can make a simple single digit counter with this project also. To make a single digit counter, remove the coil connect a push button switch between Pin 1 and power supply. Pull down the input by a 100K resistor. For each pressing of the push button the counter advances by one digit. The circuit is given in **Schematic 53.**

A word about seven segment displays.......

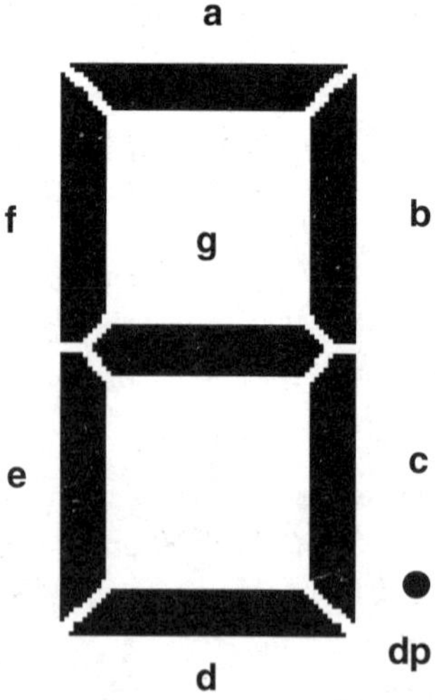

Figure 60

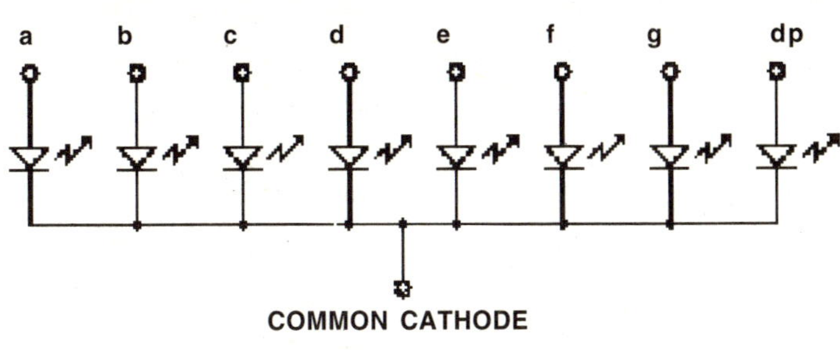

Figure 61

Look at the ***Figure 60, 61,*** **and 62.**

The segments in a seven segment display are referred to as 'a b c d e f g' with 'a' at the top and counting clockwise as shown in the figure. G segment is the center bar. These displays can be either common cathode or common anode from the way in which the LED terminals are made common. Following diagram shows how cathodes or anodes of the seven LEDs and a decimal point are made common. Either all the cathodes or all the anodes are made common to facilitate easy work up. This also allows us to enable or disable a particular digit by referencing that cathode or anode. Each type is used depending on what the driver IC requires and has no other particular advantage. Displays come in all available LED colors and you may use any color of your preference.

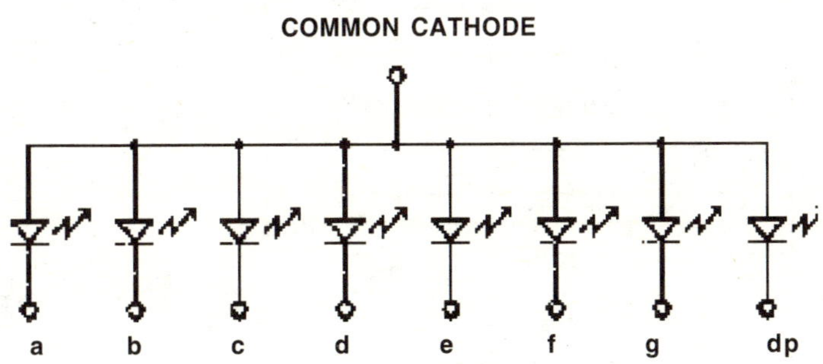

Figure 62

Construction

You can mount both IC and seven segment display digit on the same board or separately. You may need to wire jumpers to digit segment pins from the IC outputs. Think for a minute to find out how these jumpers can be minimized. 2, 3, 4, 5, 8 pins are grounded. S1 is the reset switch which is normally at low level by 100K, but when pulled up it resets all the counters to zero.

If you want to change the voltage of operation within the limits of the IC, you have to make a corresponding change in R2 value. Coil can be any hook up wire or enamel wire of a few turns.

Parts

Item	No. reqd	Description	Designation
1	1	LT543	DISP1
		(COMMON CATHODE DISPLAY)	
2	1	COIL	L1
3	1	100K	R1
4	1	1K	R2
5	1	Push to on switch	S1
6	1	CD4033	COUNTER IC

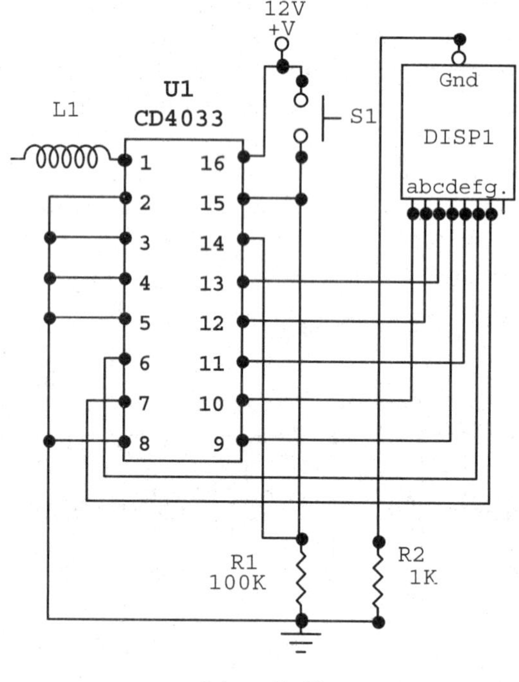

Schematic 53

Bet a Number Game

Introduction

Here we have an interesting small party game to share with your friends or with children in a school function, a kind of roulette. In stead of a roulette wheel here we have continuously changing numbers finally ending up at random.

Description

The circuit is given in **Schematic 54.** Here two CD4033 ICs are cascaded to give a two digit display. We have another CMOS quad NAND gate IC to help us with the clock pulses. Two of the gates in it are wired as an oscillator, pulses of which are fed to the clock input of U2. Carry out input of this IC is connected to the clock input of next IC.

Press S1 and U1A and U1B gates in CD 4011 start oscillating and drive the clock input of CD4033, U2. The combination continues to oscillate as long as S1 is pressed. The counter starts counting so fast that it will be difficult to keep track of the number on display. When the switch left open, the counter stops some where and that is the trick. More often than not it is not the number that was betted on. The counter will be ready for the next victim when S2 is pressed and the counter is reset.

You may use 7555 as an astable oscillator with a suitable frequency to clock CD4033.

Construction

Care is recommended as all the three are CMOS ICs. ICs and displays may be wired on a single board or on two separate boards. However we will have more flexibility in placing the display in a box if the displays are wired separately on another board. Display can now be fixed perpendicularly along with the switches. You may wish to make a large display of these digits. Then please follow the common cathode model of the display as shown in ***Figure 61.***

Try using four LEDs in series, properly fixing them as per the segment. Do not use different color LEDS. Reduce R1 and R2 to give a segment current of 10mA, but not more than that.

Parts

Item	*No. reqd*	*Description*	*Designation*
1	1	.01uF	C1
2	2	LT543 (COMMON CATHODE DISPLAYS)	DISP1, DISP2
3	2	1K	R1, R2
4	1	100K	R3
5	2	10K	R4, R5
6	2	Push to on switches	S1, S2
7	1	CD4011	U1
8	2	CD4033	U2, U3

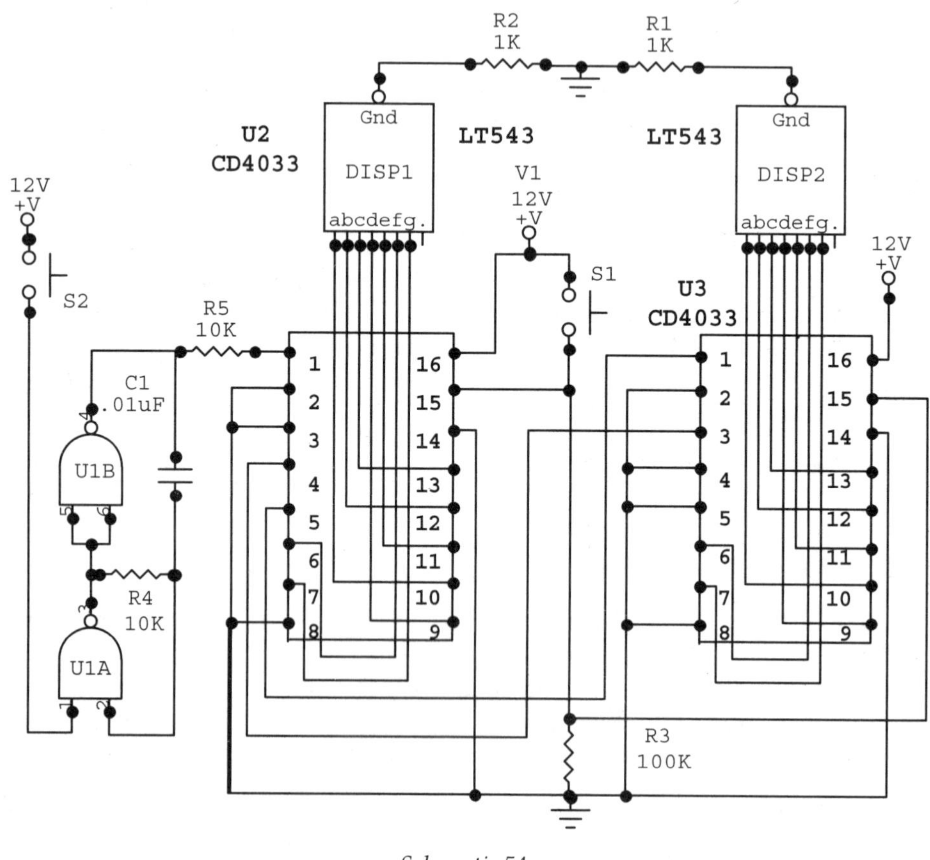

Schematic 54

Object Counter or Visitor Counter

Introduction

Now that we have built counters with CD 4033 let us go ahead with another interesting device as it is time to try our hands with very useful counter IC MM74C926. Here we have an application for counting number of persons passing through a door or a number of objects passing out or any other number of interruptions. As a totalizer, it can count up to 9999.

Description

The circuit is given in **Schematic 55.** The circuit is built around three ICs. First is our friend in need 555. It is wired as an astable multivibrator at a frequency of 36 KHz which fires two IR LEDs in series.

PIN OUT (1 = GND, 2 = VS, 3 = OUT)

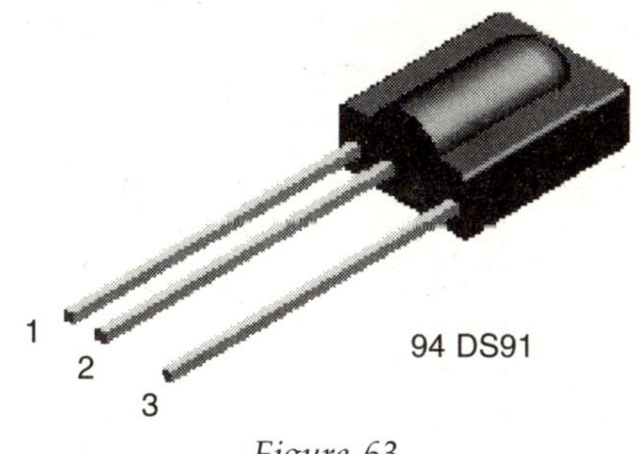

Figure 63

Next is a three terminal IR sensor, TSOP312363. Pin out and block diagram of the IC are given in Figure 63 and Figure 64 respectively.

As the name suggests it has only three terminals, Positive (2), negative (1) and output (3) terminals operating at 5V. It has Photo detector and preamplifier in one package. It has improved immunity against electrical disturbances and ambient light. Operating from 2.7V to 5.5 V it is compatible to CMOS and TTL ICs.

BLOCK DIAGRAM

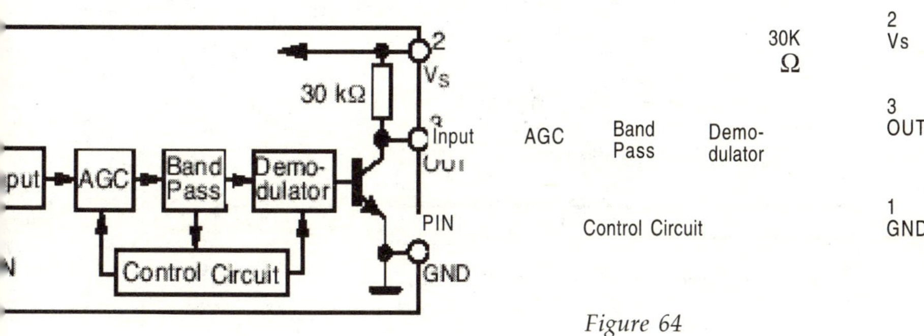

Figure 64

Now IR LED should be fixed in the direction of IR sensor as the IR pulses should reach IR sensor. If there is break in the 36 KHz stream, the sensor will send a low level signal to its output.

MM74C926 is a unique four digit decade counter with internal output latches, display drivers with multiplexed outputs. Multiplexing is accomplished by an internal free running oscillator which does not require external clock source. It has a good supply range of 3 to 6V. It has a carry out pin to cascade more number of ICs to get more number of digits on display. High level at the reset pin resets the counter to zero and reset the carry out pin to low. Pin out and the functional diagram are given along.

PIN OUT OF MM74C926

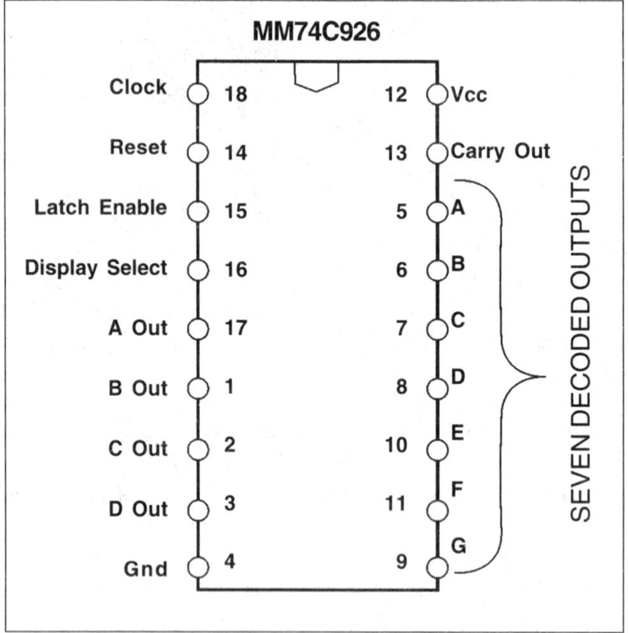

Figure 65

FUNCTIONAL DIAGRAM

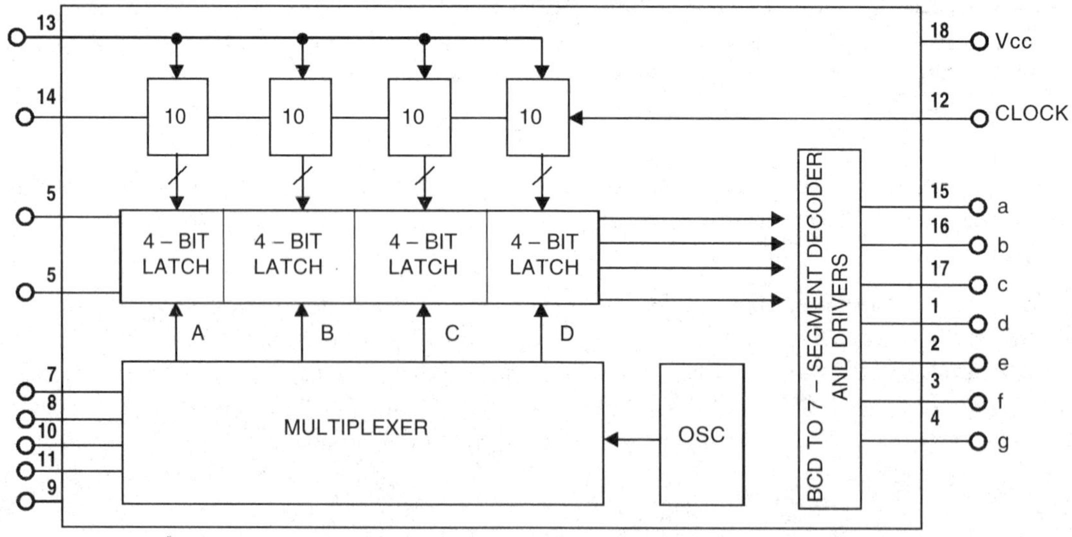

Figure 66

Multiplexing

You will notice that this circuit has multiplexed output lines. In the earlier two digit counter you could see that there are 14 output wires and two ground wires to go. For a four digit counter, they would be 32. But here in the schematic you will find that all the four digits are powered by just eleven wires. That trick is multiplexing. For a four digit display, the trick is to display the first digit, turn it off, display the second, and turn it off and so on. This is done so fast that our eye will be deceived into thinking that all the digits are illuminated at the same time. We are actually taking advantage of a characteristic of our eye known as 'persistence of vision.' Eye treats any flickering above 30Hz as a constant and that does the trick in our televisions and computer screens. It is as if God has blessed us in advance with this faculty for the modern man.

Now keep the IR LEDs of 555 at one side and the IR sensor at the other side. Direct the light of LEDs on to the IR sensor. Keep the distance depending on the sensitivity of both. As long as there is a stream of 36 KHz passing through both nothing happens. But of it is interrupted by a person or object, sensor clocks a low level signal to 74926 and the counter advances.

Construction

I am not exactly a proponent of Veroboards; but printed circuit boards exclusively made for these ICs are generally not available. Hence a general hobbyist or student may have to fall back on Veroboards. We need two boards one for 7555 and other for the main circuit. 7555 can work as low as 3V. Hence 3V battery power can be used for the portability but if the range of IR LEDs suffers, you may have to use higher voltages. But change the current limiting resistor suitably.

All the three are CMOS ICs and MM74C926 is rather expensive. Hence care is recommended. ICs and displays may be wired on a single board or on two separate boards depending on the viewing necessity. However if you wish to place displays perpendicular to the main board for better display angle, the displays can be wired separately on another board. If you wish to make a large display of these digits, please follow the common cathode model of the display as shown in ***Figure 61.*** Try using four LEDs in series for each segment. Do not use different color LEDS. Any common cathode displays can be used of the same color.

Parts

Item	No. reqd	Description	Designation
1	1	1kPF	C1
2	1	.1uF	C2
3	1	.01uF	C3
4	4	LT543 (COMMON CATHODE DISPLAYS)	DISP1, DISP2, DISP3, DISP4
5	7	100 Ohms	R2, R3, R4, R5, R6, R7, R8
6	4	1.8K	R9, R10, R11, R12
7	3	10K	R1, R13, R15
8	1	330	R16
9	1	15K	R14
10	4	BEL187	Q1, Q2, Q3, Q4
12	2	IR LEDs	D1, D2
13	1	TSOP312363	IR SENSOR IC
14	1	MM74C926	COUNTER IC
15	1	7555	TIMER IC
16	1	Push to on switch	S1

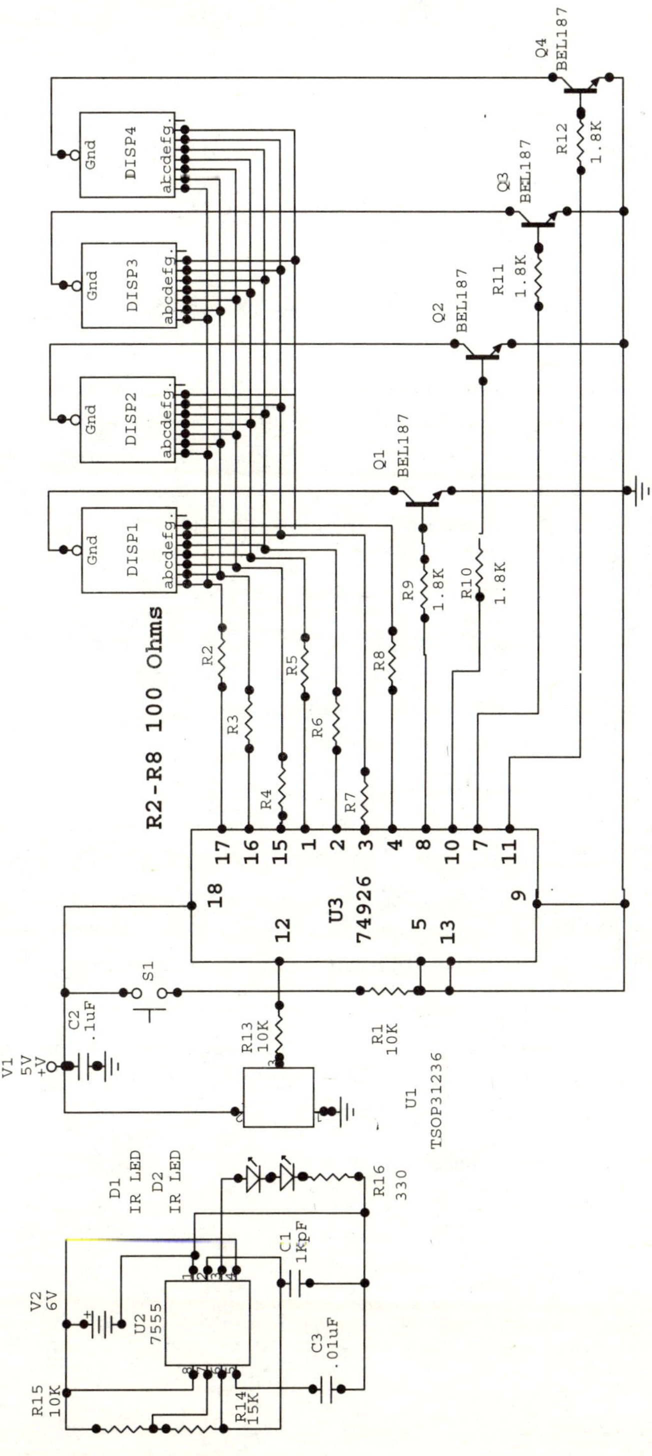

Schematic 55

Digital Clock

LM 8560, a stand alone digital clock chip boasts of some very good features like wide supply range (7.5 volts to 14.0 volts), 12-hour / 24-hour display format, on chip alarm output (900Hz tone), 50 / 60Hz input selection, on chip battery back up oscillator. For setting the clock IC fast forwards hours and minutes by push button switches. Snooze for alarm, sleep timer (maximum of 59 minutes or 1 hour 59 minutes) which can be used to drive radio or a relay are also available.

This IC has equivalents like Texas Instruments IC, TMS 3450, FORTECH make, BT 8560, and BEL MICROSYSTEMS make, IL 8560.

AC frequency (50/60 Hz selectable) available from the secondary of the transformer can be used as a basic frequency into the chip for its operation as it has an on-chip oscillator to back up the clock in case of power failure. However as the mains frequency in many countries is not that stable for an accurate clock, a crystal oscillator and divider IC (CD4541) is used here. Normal common cathode displays are not suitable here as this IC drives duplex display. But this timeshared duplexing reduces the number of display pins to 14. The chip has only 28 pins, compared to 40 pins of good old TMS 1943 or LM 5387 versions.

Pin out of the IC, LM8560 is given in ***Figure 67*** and block diagram in ***Figure 68.*** . The maximum permissible display current is 18mA and the device consumption is 0.7W. The clock displays real time, alarm time, sleep, and seconds selectable through two SPST switches. Hours set, minutes set, alarm off, snooze are operated by push button switches.

Basic frequency for the clock is taken from the oscillator, divider IC, CD 4541. Pin out and description of the IC is presented elsewhere in the book. In the present schematic, oscillator with a crystal of 3.2768 MHz is employed for more accuracy. With both pins 12 and 13 (A&B) held high, crystal frequency is divided by 65536 times and the output is available at Pin 8 as 50Hz. This is directly fed to the 50 Hz input of LM8560 at pin 25 as reference frequency. The internal circuitry in the clock divides this into minutes and hours and outputs appropriate drivers.

This frequency is also used for duplexing the display. The signals are fed to the cathode 1 of the display through transistors Q1 and after inversion by Q2 and Q3 to cathode 2.

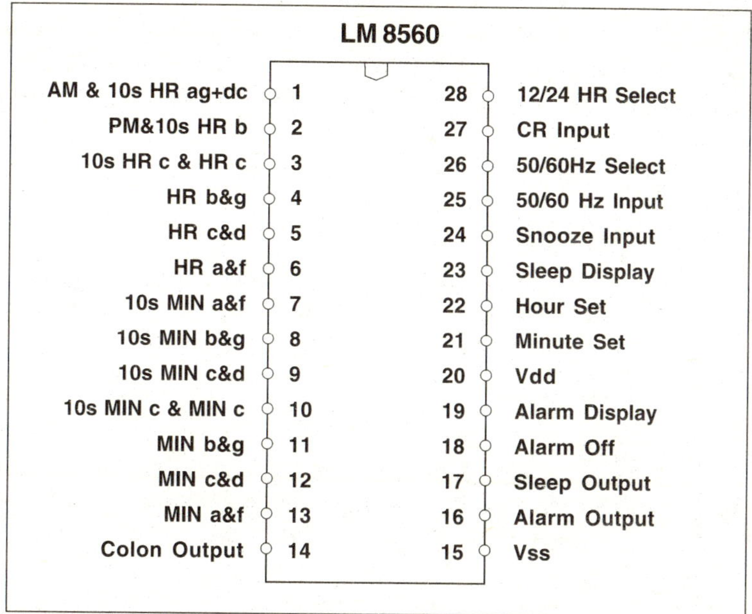

Figure 67

BLOCK DIAGRAM

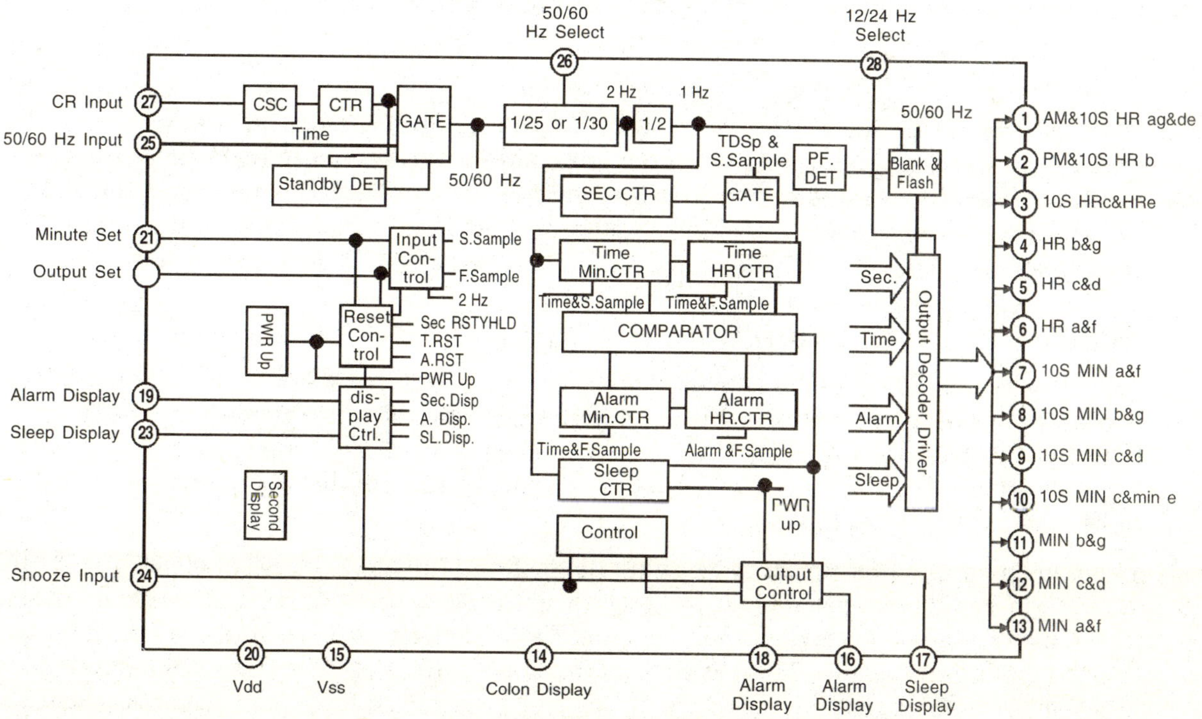

Figure 68

Setting and Display Modes

The block diagram of the clock is shown in ***Figure 68.*** Clock schematic is shown with 50 Hz operation in 12-hour mode **Schematic 56.** You have the pin out details of SP4022A a Duplex Display in ***Figure 69.***

The functions of LM8350 are managed with switches S1 and S2, (SPST switches), which furnish all the four display functions such as real time, alarm time, and sleep timer and seconds display.

If the pin 28 is left unconnected the clock goes into 12-hour mode and displays PM LED also and by pulling it to Vss the clock shifts to 24-hour format. For 50 Hz operation, connect 50/60Hz selection pin (26) to Vss.

Display Management

The circuit uses two SPST switches to display real time, alarm time, sleep, and seconds. Four push button switches are used for setting hours, minutes, and for making alarm off and snooze functions.

In case of power failure, all the display drivers turn off. However in the present schematic, CD 4541 continues to feed the input of the clock and the clock maintains time with the back up battery. When the power resumes, the display will light up and clock shows the actual time. If there is no back up, when the power returns, all the on-segments in the display blink indicating power fail and show illegal time. It can be reset by hour set or minute set switch.

Normal Time Display and Setting

With both S1 and S2 switches open, the display shows real time in hours and minutes. Pressing the hour set switch S4 pulls pin 22 to Vss and the clock increments hours at 2Hz rate. Pressing the minute set switch S5 pulls pin 21 to Vss and the clock increments minutes at 2Hz rate. Pressing both switches at the same time increments both hours and minutes at the same rate simultaneously.

Alarm Time and Operation

Switch on S2 and pull up Pin 19 to display alarm time. Now alarm time is set in a much similar way as the normal display with hours and minutes set switches. (Change alarm time by using hour set switch S4 and minute set switch S5 just like normal time display.) But now if both switches are pressed together, the alarm time is reset to 12:00 in the 12-hour mode or 00:00 in 24-hour mode.

When the set alarm time synchronizes with the real time, an alarm signal of 900Hz gated at 2Hz at Pin 16 will be available. This signal drives a piezo speaker for 1 hour 59 minutes through R18 and Q4. Press S6 to pull pin 18 to Vss and switch off the alarm. If switch S3 is pressed pulling snooze pin (24) to Vss, the alarm repeats after 8 or 9 minutes, while the alarm time is active. 900 Hz alarm signal can be turned to DC signal by a simple low pass filter, which can drive a relay or switch a radio circuit at the alarm time.

Sleep Timer Display and Operation

Sleep counter can be used to drive a relay or switch a radio or any other circuit up to 1 hour 59 minutes. Sleep timer is activated with the switch S1 at Pin 23 pulling it to Vss. It is immediately reset to 00:59 minutes and the down count starts. This fires a relay through R19 and Q5 and it will be on for the time set. But if both sleep and hour set pins are pulled to Vss the display sets to 1:59 minutes.

If minutes set switch S5 is pressed, the sleep display decrements at 2Hz rate and required sleep time can be set. Counter down counts until 00: 00 hours when its output goes low. If the snooze pin is pulled up the sleep timer goes down to 00:00 hours.

Second Display

With both alarm and sleep display switches S1, S2 closed and pulling Pins 23 and 19 to Vss respectively the clock enters seconds display. You can hold the clock, by pressing the minute switch. If both hours and minutes switches are pressed simultaneously time now resets the time to 12:00 in the 12-hour mode or 00:00 in 24-hour mode.

Parts

Item	*No. reqd*	*Description*	*Designation*
1	1	Piezo Buzzer	BZ1
2	2	10pF	C1, C2
3	1	1000uF	C3
4	1	BRIDGE	D1
5	1	1N4003	D2
6	4	BC147B	Q1, Q2, Q4, Q5
7	1	BC157B	Q3
8	14	390	R1, R2, R3, R4, R5, R6, R7, R8, R9, R10, R11, R12, R13, R14,
9	1	1M	R15
10	2	10K	R16, R17
11	2	2.2K	R18, R19
12	1	4.7K	R20
13	1	12V 300Ohms	RLY1
14	2	SPST	S1, S2,
15	4	Push to on switch	S3, S4, S5, S6
16	1	230V/12V Transformer	T1
17	1	12V Battery	V2
18	1	3.2768MHZ	XTAL1
19	1	Peizo buzzer	BZ 1
20	1	LM8560	
21	1	CD4541	

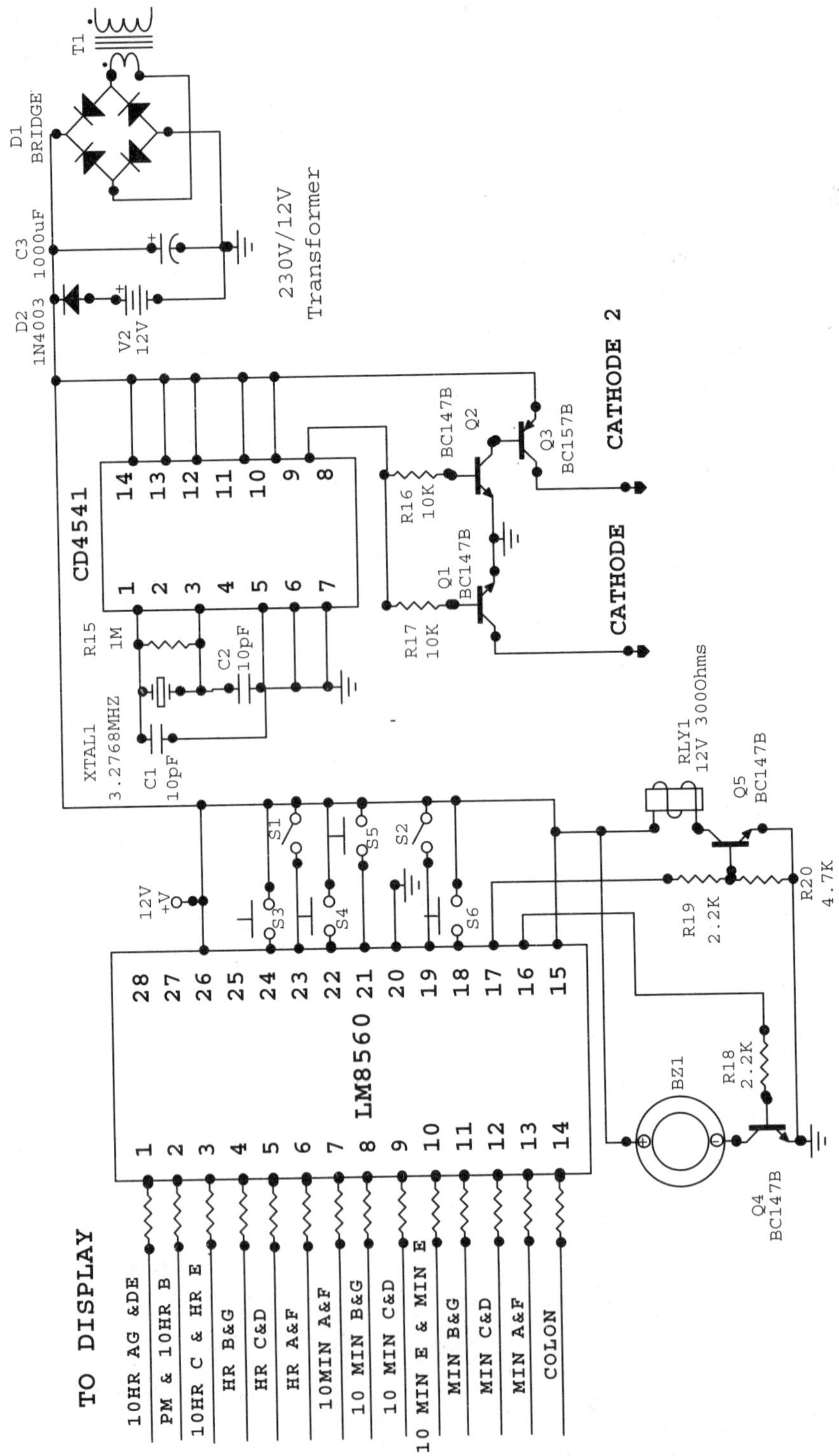

Schematic 56

DISPLAY PIN OUT

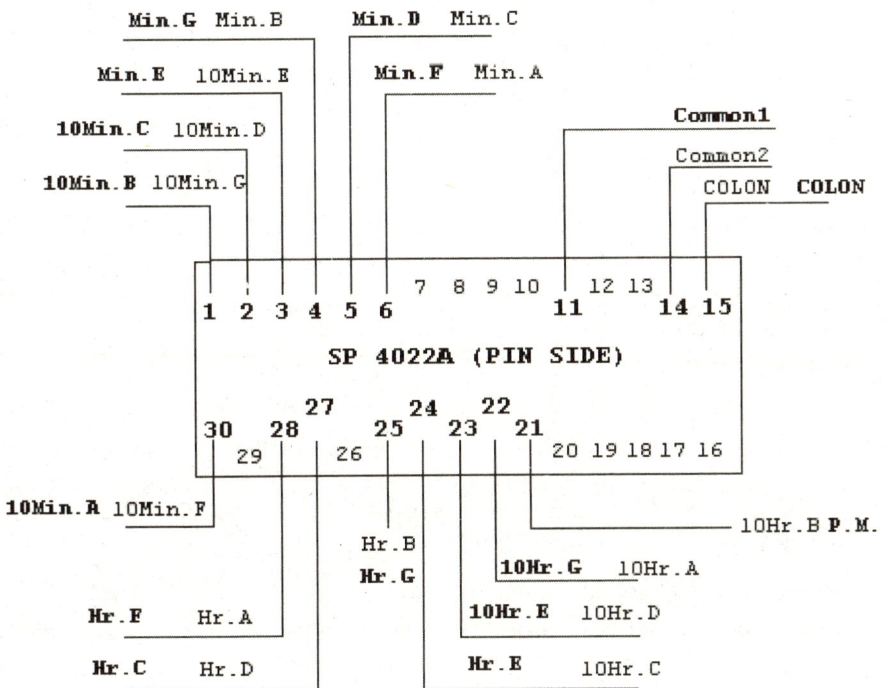

Construction

Schematic look a bit complicated but is quite easy. Power supply is simple and straightforward. 230V AC is transformed by a 500mA transformer T1 to 12V and rectified by bridge D1. 1000 µf / 25V filter capacitor is more than adequate. The clock should keep working always, with or without display, so that we can look for an accurate time always. Hence battery back up is provided by 12V batteries through D4 diode. You may use any kind of cells, even rechargeable cells, though charging circuit is not included.

Hard wire the display with long enough wires so that it can be placed conveniently for a visual display. Varieties of plastic boxes are available now. Pick up which can accommodate transformer, clock boards, relays, etc. If you do not wish to have relays, you may do away with it. If you wish to make jumbo displays, take a box sufficient for it.

Display Drive and Jumbo Display

Normal seven-segment displays cannot be used with this IC, as it supports only duplex display. Instead of making common connection of all the cathodes of the seven segments from all the four digits, here all the cathodes are divided in two. Cathodes are driven by alternate cycles and anodes are driven by duplexing. Some of such displays are SL 1498T, TLR4262, LT637 and SP4022A. You already have the pin output details of SP4022A and LM8560. Follow them carefully when interconnecting.

Jumbo displays with four LEDs in series for each segment can be built. Individual segments were connected in such a way to make two common cathodes for all the four digits. You may take tips from the Counters section of this book about the displays.

Let us take for instance, Pin 11 driving G and B segments of minute digit. Four LEDs were connected in series at the said G segment and another 4 LEDs were also connected in series at the B segment of the same digit. Now both the anodes are made common and connected to Pin11 through 390 ohms resistor. Now cathodes of all LEDs in G segment were made common and connected to Common Cathode1 and similarly cathodes of all LEDs in B segment were made common and connected to Common Cathode 2.

Now take Pin 12 driving D and C segments of minute digit. Four LEDs were connected in series at the said D segment and another 4 LEDs were also connected in series at the C segment of the same digit. Now both the anodes are made common and connected to Pin 12 through 390 ohms resistor. Now cathodes of all LEDs in D segment were made common and connected to Common Cathode1 and similarly cathodes of all LEDs in C segment were made common and connected to Common Cathode 2.

This way, the segments are built up and cathodes are made common and connected to common 1 or 2. Two anodes each as described in Table below are connected together to the respective segment drive pins through current limiting resistors.

You cannot use different colored LEDs. If gains of Q1, Q2 and Q3 are not matched, there may be variation in the brightness of the segments.

Pin No.	Cathode 1	Cathode 2	Pin No	Cathode 1	Cathode 2
1	10 Hour **A&D**	10 Hour **G&E**	2	**PM**	10 Hour **B**
3	Hour **E**	10 Hour **C**	4	Hour **G**	Hour **B**
5	Hour **D**	Hour **C**	6	Hour **F**	Hour **A**
7	10 Min **A**	0Min **F**	8	10Min **B**	10Min **G**
9	10 Min **C**	10Min **D**	10	10Min **E**	Min **E**
11	Min **G**	Min **B**	12	Min **D**	Min **C**
13	Min **F**	Min **A**	14	**COLON**	

Let the clock tick now!

High Low Voltage Cut Off

Introduction

Voltage fluctuations are commonplace in most parts. Costly equipment is damaged due to unforeseen variations in voltage. As a general rule, most of the modern electronic equipment can tolerate much lower voltages but not the high voltages. On the other hand, equipment like refrigerators and air conditioners which run with motors cannot tolerate both high and low voltages. Here is a circuit that trips the power when the voltage goes high or low beyond certain limits. It warns when the power is disconnected and is built around inexpensive components.

Description

The circuit is shown in **Schematic 57.** Good old full wave power supply with 12V is used now. Reduce the filter capacitor to 220 mfd. Green LED D2 gives power supply indication.

When mains voltage is within limits, DC voltage at the zener diode (D3) cannot make it conduct. Hence transistor Q1 is cut off. But DC voltage at zener diode D4 is greater than break over voltage of 6.8V and hence transistor Q2 conducts. So the relay RLY1 gets energised and powers the equipment. Now if the mains voltage increases, zener diode (D3) conducts. Hence transistor Q1 also conducts. This pulls Q2 into off state. The relay de-energises cutting off the power. If the voltage drops, Q2 any way will not conduct and the relay will be de-energised.

In both the cases, when Q2 cannot conduct, its collector potential will be high. It is transferred to the piezo buzzer through diode D6, which warns that the power supply has gone bad. Red LED D7 also lights up indicating disruption of power to the connected equipment.

Relay contacts should be capable of handling the current requirements of the equipment to be powered.

Construction

Main power supply is the old one. All the other components are discrete. They can be soldered on a piece of vero board. Adjustment of the presets requires a little care. Take an auto transformer or a variac. It is a device which can continually vary the voltage fed to the input of the circuit.

Connect it to T1 transformer and increase the voltage to about 260 volts. Adjust the preset VR1 so that Q1 goes into on state. Reduce the voltage and see if the transistor goes off again. Then reduce the voltage to about 200 volts and adjust preset VR2. Carefully adjust until Q2 goes into non-conduction. Return the voltage to normal and check if the transistor has reverted to its original position.

The adjustments can also be made with adjustable DC power supply with LM 317. Do not power the circuit now from the mains transformer. But connect 12V regulated supply at the positive and negative rails. Check if the circuit is behaving normally. Increase the voltage correspondingly and adjust VR1. Similarly decrease the voltage and adjust VR2. Check the conditions. 1V DC can approximately indirectly simulate mains voltage of 20V.

Parts

Item	*No. reqd*	*Description*	*Designation*
1	1	Piezo Buzzer	BZ1
2	1	220uF	C1
3	1	IN4003 X 4	D1
4	1	GREEN LED	D2
5	2	6.8V Zener Diode	D3, D4
6	2	IN4003	D5, D6
7	1	RED LED	D7
8	2	BEL187	Q1, Q2
9	4	1k	R1, R2, R3, R4
10	1	12V 300 OHMS Relay	RLY1
11	1	230V/6-0-6V Transformer	T1
12	2	5K Presets	VR1, VR2

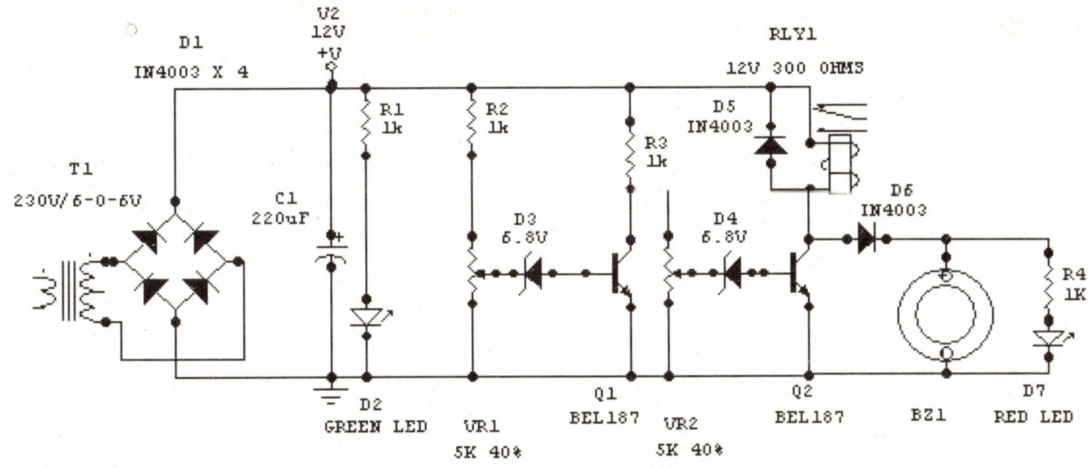

Schematic 57

Super Simple Twilight Switch

Introduction

A light dependent resistor, one resistor, one preset resistor, a capacitor, one triac and diac combination is all that makes the porch light to go on at the nightfall and switch off at dawn. Start if you have them. Circuit operates at mains voltages of 230V and please be careful that it does.

A super simple twilight switch circuit, which I made out of components from the junk box, which I am sure every hobbyist has his handful. No ICs, no programs and made with parts that can be fitted in the light switch box in just one afternoon.

Description

Schematic 58 shows a twilight switch circuit. The circuit directly operates on mains. LDR, R1, VR1 preset are connected in series across AC mains. Middle terminal of VR1 is connected to the diac, which in turn is connected to Triac Q1. If there is enough falling on the LDR, the resistance combination consisting of LDR, VR1, R1 cannot overcome the break over voltage of Diac. Hence triac cannot fire. Adjust R2, sensitivity control and trigger the triac at the required light threshold. Triac used is BT 136, rated at 600V at 4 amps. Diac DB3 is rated at 33 V break over voltage.

I, of course used a quadrac (Q6004), which is rated similarly. It is nothing but a triac with diac built into a single package. If you have an old but working electronic fan control, some of these parts can be retrieved from there.

Construction

Any triac rated for mains operation will do well. It is essential to keep the LDR away from disturbing lights and from the own light of the circuit, which may false trigger the circuit. If you are using the IC Vero board, it is good to remove one track and solder components on alternate tracks. Remove all solder flux and make the board dry before powering it. High voltages and high frequencies do not like bad housekeeping. The circuit can work on incandescent lights only.

Parts

Item	No. reqd	Description	Designation
1	1	.1uF/600V	C1
2	1	DIAC	D1
3	1	Light Bulb	L1
4	1	BT136	Q1
5	1	47k	R1
6	1	47K var	VR1
7	1	LDR	R4

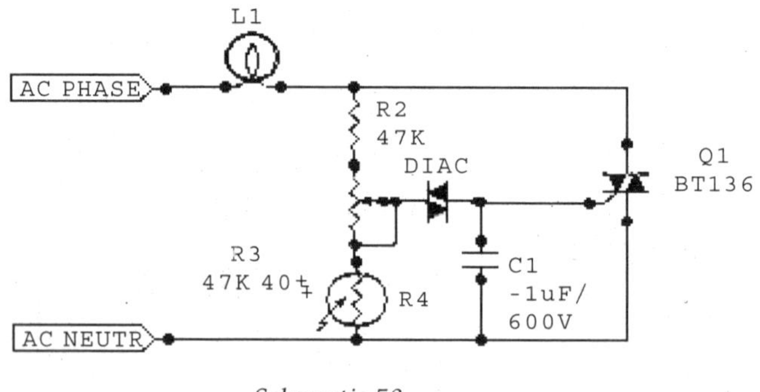

Schematic 58

Tips and Tricks

Think

First think what you want to do. Write points on a paper. Write component list. Check in your boxes if you already have them. If you have them, segregate and tick off in your list. It is not a good idea to use very old electrolytics. Some components, like resistors can be safely and easily removed from old boards. But it takes a lot of skill and care to desolder ICs from the boards and still make them work. Buy the components which are not available with you. Do not compromise on quality of components, especially capacitors, electrolytics in particular. If you are new, buy IC bases also.

Make your board ready

If you are using a Veroboard, think and think of the component layout, i.e., how you are going to place components. Try the shortest way to interconnect parts without jumpers. Plan the components in such a way that the printed board tracks make interconnections. Cut the track only where necessary. Cut small line, but make sure that all the copper is gone. Some people actually drill out the track at the holes. This makes the board weak. A small copper left in the track can create havoc and similarly you may have to bridge the cut track if a mistake is made or a change is required. Experience will tell you that it is extremely difficult to bridge a broken track with solder alone. Clean the track with a blade or knife just before soldering a component. Shining copper makes all the difference in soldering.

Make your parts ready.

Check if you have all the parts. Clean and scrap the wire leads of the parts. It will be virtually impossible to solder a dirty and unclean part. Do not cut the leads short. It will be easy to insert parts in the holes with all the lead, solder and then cut them. In any case do not cut the leads of semiconductors like diodes, transistors, ICs. Many times leads will help in heat sinking. Use sleeves for preventing leads from shorting out.

Solder small parts first.

It is better to solder small parts first. Solder resistor, capacitors, diodes, transistors, large capacitors and ICs in that order. However some times it is convenient to follow the order of the circuit and lay them. Preferably keep the semiconductors to the last, so that they are not unnecessarily heated up or abused. Use a good soldering iron with low leakage element or grounded iron. A street smart idea is to remove the soldering iron from mains and solder static sensitive parts like CMOS ICs. It is little time consuming but safe. Read and follow instructions on handling CMOS, TTL and linear ICs like Audio ICs.

Use heat sinks

It is better to use a small clip as a heat sink on the semi conductor leads, because they are heat sensitive. Even pliers will help. Solder as fast as possible while handling these.

Check and recheck

Check all the parts once again if they are properly soldered, whether they are inserted right way in. Use bases for the ICs. Check if they are inserted properly. Even the experienced hobbyist did make the mistake of inserting ICs wrong way. Check for bad solder joints. If only two reasons are to be assigned for project failures, they are bad solder joints and wrong placement of components. Check and recheck. Test the tracks for shorts and for proper interconnections. Try not to re-solder on the boards with CMOS ICs already soldered or use the method suggested above.

Test

If possible test the circuit on a breadboard particularly if the parts are costly. There will be errors in the circuits printed in the magazines or sourced from websites. I often advised people to wait for a month before trying any new circuit from the latest magazine, just to know from the letters section if any errors are pointed out.

Try

Power up and look carefully for the first few seconds for obvious signs. Check if the applied voltage is stable. Check if the speaker squeaks or a LED dips. If the voltage falls appreciably, switch off and check for possible shorts. Most possibly, there will be a short in the cut tracks, or a wrong insertion. In spite of all the care, there will be Murphy's Laws at work and you may end up with a failed part, costly or not so easily available. Do not loose heart. Edison made 2000 attempts before making a successful light bulb.

Enjoy

But if every thing is fine, enjoy. There is nothing to beat the pleasure we derive from the successful project made by us alone. Then.............

The next one....

Do not stop. Start working on the next one.

Soldering

Soldering is a skill that can be acquired easily but it is true that I used to feel jealous of the people who made good soldering joints. It is technique where metal with low melting point such as lead or lead alloy is used to join the surfaces with a low heat or a kind of hot glue.

Beginners would be generally apprehensive about the component damage during soldering operations. Components such as transistors, diodes and ICs do get damaged due to excessive heat. This can be avoided by learning sound soldering techniques. As always, patience makes a man perfect. A little care and training will make the task easy. Follow these simple steps and learn those small tips and tricks.

Soldering iron and solder

35W watts soldering iron is good if you are using the good old nichrome element iron but a 25W iron will suffice if it has ceramic element. Ceramic element iron better protected for stray leakage. Do not use higher wattage iron and never use the so called soldering guns.

Rosin cored solder of 60% lead and 40% tin combination is suitable for electronics application. 70-30 combination also will work well. Present soldering wires have flux in their core. Take the wire suitable for your application. Additional flux also is used depending on the necessity. But remember extra flux can create havoc particularly in CMOS and audio applications. Never use lead with acid core flux as this will damage the tracks, leads and other components.

Please be careful about the harmful fumes released when soldering, because of the flux and the lead. Actually the world is turning towards leadless soldering. Please mind the soldering iron as it is hot and once I almost lost my eye when it slipped out of hand.

Tinning the iron

The tip of the soldering iron should be tinned as this will greatly help in doing the job fast and proper. It actually improves the heat transfer. Clean the tip with a fine emery or steel wool and apply soldering lead and spread it over the surface, while it is hot. If necessary add a little additional flux.

Preparation

All surfaces to be soldered must be clean. Scrap the leads with blunt knife or some thing like that. Similarly clean printed board tracks also. It helps to make the track on the board shine at the point of soldering. You can never get a good and low resistance joint with bad surfaces. You will only be dumping lead and flux.

Place components

Next step is to place the components. Insert the parts with all their leads. Now it will be easy to keep parts in place after bending the extra leads at an angle at the board. Cut them close after successful soldering. However one can get a feeling that God must have given us three hands at least while soldering. Make doubly sure that the parts are placed right way in, because you need a skill to solder but an art is required to desolder them.

Heat of the moment

Touch the hot iron on the component lead and the board with just enough pressure to transfer the heat. When the joint is sufficiently hot and bring the solder wire onto it. Solder should touch the parts, not the iron. Flux in the wire should be enough but you may add a judicious little more. It should take about two to three seconds for the joint to make but larger wires and pads may require a little more time. If every thing is fine and hot, solder flows beautifully around the joint. Remove the solder and then the iron. Keep the joint steady and still for a few seconds. Make a move, you will end up with a cold joint and if it is not hot enough, it will be a dry joint. Both make bad electrical connection and can create intermittent contacts making fault finding difficult.

Housekeeping

After all the joints are made, do not forget to clean up excessive flux. Check once again for solder bridges.

Repair

Bad joints are characterized by dull grainy structure. Remove the old solder by reheating it and pushing the molten lead aside. Check for component and board cleanliness. Add new lead and resolder. However if you are sure that the parts are clean, you may reheat the solder, add flux. It should make the joint. Steady hand is required as always.

Tips and tricks

- Solder tips should be clean as well as the components and printed circuit pads. Do not compromise on that.
- They should be tinned for an easy job.
- Tinning the parts also greatly helps in fast soldering
- Steady hand and mind keeps the parts in place and avoids cold joints.
- Use heat sinks while soldering sensitive components like ICs. A pair of pliers or even good crocodile clips will do well.

- Do not keep heating the parts. If you cannot solder in five or ten seconds, there is some thing wrong. Prolonged heat is the shortcut to damaged ICs. Practice on less sensitive parts.
- Use right type of iron and the tip. 25W to 35 W irons should be alright but you may need a more powerful iron to solder heavy metals. Do not use solder guns.
- Solder small parts like resistors, jumper wires, diodes, and small capacitors and then go to large ones like electrolytic capacitors
- Recheck the joints after they are made. If in doubt use ohmmeter, for checking shorts and open circuits.

Desoldering

Soldering is a skill but desoldering is an art. You need the soldering iron of the same wattage and make. Additionally, you will need a solder sucker or desoldering wick. Most often both are used.

Solder sucker is a kind of a vacuum pump. It has a spring loaded plunger, and a button to release it. There is a small Teflon nozzle on the bottom side. The plunger is pushed down by a knob at the top. Heat the solder to be removed until it melts. Position the nozzle over the molten solder and press the release the button. By pressing the button, the plunger moves back, creating a vacuum at the Teflon nozzle. It sucks the solder up.

Solder wick is a braided copper wire and with a little skill you may even use the shield from the shield wire used for audio cables. Needless to say the wire should be clean and shining. Melt the solder to be removed and bring the wick on to it. Solder will flow on to the wick. You may have to take the help of gravity.

Many technicians use both techniques. Remove most of the solder lead with sucker and finish it with wick. Wicks cannot be reused. But soldering pumps need to be occasionally cleaned. Remove the nozzle. Take out the accumulated solder and fix back the nozzle. You can reuse this lead using normal soldering flux.

Fine if you are removing definitely damaged part. But if you are trying to remove misplaced or wrongly positioned part, please take all the precautions not to overheat or break the part.

Tips and tricks for soldering can generally be used for desoldering as well.

ICs and Care in Handling

The precautions given below are not in meant for prospective hobbyist to drive him away but to keep him informed of the pitfalls. Earlier in the book, it was often repeated about the care of CMOS ICs, but most of the modern CMOS ICs are extremely rugged and they can tolerate most of our mistakes. So are other ICs. They are in fact rugged and quite tolerant. Foregoing is notes on certain precautions to be taken while handling ICs, but it is not meant make you run away from these ICs.

Integrated circuits are broadly divided into linear and digital versions. Audio, video and such other ICs are treated as linear ICs, where there is finite output for the input. Digital ICs are further basically divided as CMOS and TTL versions. 4000 series are generally CMOS versions and 74 series belongs to TTL family. Basic TTL ICs are on their way out with the advent of 74 HC, 74HCT, 74LS, etc., versions.

Main advantages of ICs have in their protection circuitry, ease of design, relative simplicity of external circuitry and components, quick assembly and high reliability. They operate reasonably well in wide supply range.

Do not use ICs in the vicinity of their maximum rating, because even a slight upward variation can damage IC. Always allow a margin unless the power supplies are stringently regulated.

Avoid shorts between pins by solder or by stray lead or copper. When applying power or before inserting IC onto the board, always make sure that there is no short across by any of the pins. This is particularly true in case of Veroboard. Leftover flux is sure to create trouble.

While inserting IC or soldering it onto the board, make doubly sure that the pins are right way in. Wrong insertion may lead IC to something unusual, thereby leading to deterioration or failure of IC. If you have learnt the art of soldering, you will know that it is much more difficult to remove a soldered IC. More often than not, it will be damaged.

If you wish to make a printed board, follow the sample as suggested by manufacturer. No feedback loop must be formed between input and output, particularly for audio and linear ICs.

Radiator fins on audio ICs must be connected to ground. Then heat sink, thermally conductive material such as copper or aluminum is mounted on IC to remove the heat as it develops. It must be of enough size. Solder IC pins only after the heat sink is fixed. No foreign particles should exist between heat sink and radiator fin, as this will greatly impair the performance of the IC. Use heat sinking compound between the surfaces.

ICs dissipate heat as they dissipate more and more power at more and more voltages. The heat must be removed continuously such that IC operates at rated temperature. Failure to do so will result in thermal runaway. If the IC is well protected, output power will fall to a safe area and it will operate at a lower power. If it cannot it will eventually fail and pack up. Copper foil area is also made as large as possible in the vicinity of IC to dissipate more heat.

Solder pins fast. If you cannot solder in less than 10 seconds, stop, do some thing else and come back.

Do not draw more than rated power from outputs. Output loads must be never be increased, i.e., output resistances and impedances must never be reduced without consideration to rated power.

Some plug jacks used for connected to the external speakers can make short between pins while inserting. These can damage the ICs.

CMOS ICs

CMOS ICs (*complementary-symmetry metal-oxide semiconductor)* are widely used in digital electronic circuit because of their low power dissipation, wide supply range, and good noise immunity. They can source or sink about 5 to 10mA, enough to light up an LED, but for higher currents, use a transistor. Power consumption of the IC itself is a few microwatts. But they have a maximum frequency response of about 1MHz.

However certain care should be taken as they are susceptible to failure due to static electricity and other characteristics peculiar to them. Please note that 74 HC, 74HCT versions of TTL ICs fall into the category of CMOS version.

Because of the high input impedance, these devices are prone to static. Static charge generally generates due to friction between insulating materials. There are so many plastic materials in every day use and one can easily get charged by them without ever knowing. For instance, by a brisk walk across synthetic carpet, one can easily accumulate charge of several hundred volts. Sitting and moving about in plastic chairs is another example. This voltage can kill CMOS ICs. While in most of the modern ICs, the input gates are protected by a diode combination, still extreme care should be taken while handling these. Do not touch the pins with bare hands.

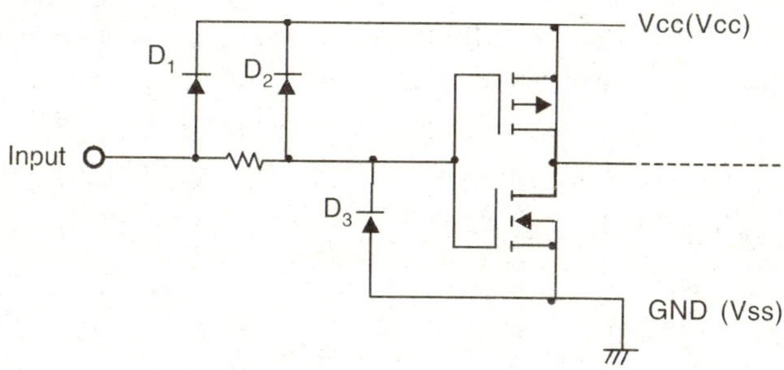

CMOS ICs are supplied in some form of conductive packing, like aluminum foil, or black (carbon filled) plastic foam or anti static bags and tubes. A wise way is to use IC holders and to insert the devices only after the layout has been checked and construction is complete and. They should not be inserted while power is on. If there are power supply capacitors on board, discharge them before inserting or removing these ICs from the boards.

Place such an IC or circuit board containing static-sensitive device on a conductive surface, such as a sheet of aluminum foil rather than on a plastic surface. Avoid touching the leads with bare hands or with metal parts. The level of static electricity in your body is enough to destroy these ICs. While an earthed mat and earthed wrist wrap are suggested while handling these devices, for a normal hobbyist handling routine CMOS ICs, they are not necessary.

Use a well insulated and grounded soldering iron. Select an iron with low leakage current. Solder in less than 10 seconds. Needless to say that the device should be inserted the right way round.

CMOS IC inputs have very high impedance and are voltage-driven. Hence, if unused input pins are left open, they are charged to unstable input voltages making the corresponding outputs also unstable. Malfunction and eventual failure due to high static build up can occur. To avoid this, the unused input pins must be connected to GND(VSS) or VDD(VCC). Leave the output pin open.

When the input voltage drops below the GND level, forward current flows in the protection diode D3, prevents a high negative voltage from being applied to the gates. When the input becomes higher than VDD, forward current flows in D1 and D2, preventing a high positive voltage from being applied to the gates. The input protection diodes are designed so that current flows through them when necessary. Take care so that current does not flow through them during normal operation.

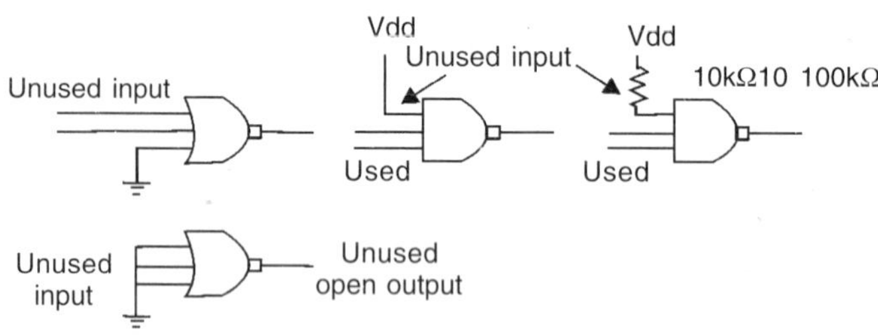

Do not connect the output pin to ground or power supply as almost all these ICs are not short circuit protected or current limited. If a capacitor is used at the output connected to ground or power supply, be careful that the charging current is not beyond output limits.

TTL ICs

These are known as the transistor-transistor logic family ICs. Original TTL ICs were numbered from 7400 or 5400 which are more or less obsolete as we have now sub families depending on the circuitry used. However end number still hold good generally.

These are better tolerant of static discharges. They (74, 74LS, 74 HCT) need strict 5V power supply and cannot tolerate higher voltages or surges. Hence they are not suitable direct battery operation of 6V.

74LS (Low-power Schottky) family uses TTL logic but is fast. It consumes more power than later families. They need 5V for operation. They can sink about 16mA, enough to light up an LED, but for higher currents, use a transistor. They can source only 2mA. Power consumption of the IC itself is a few milliwatts. They have a maximum frequency response of about 35MHz.

74HC family uses High-speed CMOS circuitry, and combines the speed of TTL with the very low power consumption of 4000 series. They are CMOS ICs with pin to pin compatible to older 74LS family. They can operate between 2 to 6V. They can source or sink about 20mA, enough to light up an LED, but for higher currents, use a transistor. Power consumption of the IC itself is a few microwatts. They have a maximum frequency response of about 25MHz.

74HCT family is a special version of 74HC with 74LS TTL-compatible inputs so 74HCT can be safely mixed with 74LS in the same system. In fact 74HCT can be used as low-power direct replacements for the older 74LS ICs in most circuits. 74HCT has a lower immunity to noise, which is not a problem in many cases. They need 5V for operation. They can source or sink about 20mA, enough to light up an LED, but for higher currents, use a transistor. Power consumption of the IC itself is a few microwatts. They have a maximum frequency response of about 25MHz.

Problems and Solutions

As a hobbyist, be prepared to face problems and failure of components. Remember the old adage, "The person who makes no mistakes, makes nothing at all!"

There can be lot of problems with audio stages like intermittent faults, drop outs, squealing, motor-boating, distortion or simply lack of gain and in worst case output short. Soon after power up, check the supply voltage. If it falls abnormally, switch off and check. There could be a short some where, a solder bridge, wrong jumper, a reversed component or even shorted and abnormal output load.

Check voltage at the mid point of output transistors or at the out put pin of the IC. It should be about half the supply voltage. If it is not, probably output capacitor is bad or shorted or even the output is short. Remove the output for moment, and see if the voltage returns here or at the power supply pin. This should give a fair indication of what is wrong. In a broad generalization, there will be around 0.6 to 0.7V measured between emitter and base of bipolar transistors and few volts to power supply line volts across emitter and collector. Mark the polarity for NPN and PNP transistors.

Check for bad or reversed electrolytic capacitors on the power supply line. Old electrolytics may go out of form and will not show rated capacitance. If you are sure of all the components, replace the IC or output transistors.

The problem could also be from the power supply as badly regulated or filtered supply will drop appreciably upon load. **Always measure your supply voltages first! ICs do not like excessive voltages.** If you have a circuit diagram on hand with pin voltages, it is a great help. And a circuit diagram with pin voltages, it is real pleasure. Unfortunately most manufacturers do not care.

Lack of gain is most often caused by bad electrolytic capacitors in inter stage, boot strap and bypass couplings. Old electrolytics dry up reducing their capacity. Counterfeit capacitors have wrong values printed n them. A normal hobbyist will not have meters to check large value capacitors. Most often capacitors can be bridged with a new and known value capacitor while in service.

Then there is unfortunate hum and buzz. Very annoying but can arise about variety of reasons. Buzzing can be caused by the input wiring close to mains leads, bad power transformer, bridge rectifier and filter capacitor and wrong ground connections. Humming is most generally caused by earth loops by input, output and power supply earthings. All earths are not same and they can not be grounded at the same place.

Motor boating is caused by bad decoupling capacitors, those normally interposed between power supply lines along the circuit. This is usually a small value electrolytic (10u to 250mfd) along with a low value resistor on the power rail. This configuration actually decouples any signal crossing over from output to input along the power supply lines which can force the circuit into oscillation. A bad electrolytic here results in a typical phut—phut sound just like from the exhaust of a motor cycle. Try bridging a known good capacitor on the suspected one and find out if it reduces or removes the sound.

Two major problems faced by a hobbyist are solder bridges and hair line cracks. Hair line cracks are more difficult to trace and can be very disturbing. Measurement with a multimeter and careful look up by a magnifying glass will help. Next problem is lifted out PCB tracks or pads. Obvious reason for this is excess soldering heat. Only solution is to run a jumper along PCB track and solder from point to point.

Touch and feel the temperatures of the components particularly at the power end. **(Definitely not on the mains operated systems).** If the temperatures are abnormal and uncontrollable, switch off and check. Do not use while the problem persists. It will eventually fail.

Modern ICs pull down the power levels to safe area in case of inadequate heat sinking or wrong and excessive output loading. Please find out if the lack of output power is due to improper heat sink or excessive power handled dissipated. Proper mounting of ICs and power transistors is important as they should make positive contact with the heat sinks.

Make no mistake that fuse will protect your circuit and its components. It is just an additional necessary protection. ICs and transistors pack up before fuse blows off as they are faster than a thermal fuse. Hence do not compromise on fuse rating at least.

As a general rule do not use speakers of less impedance than prescribed in the datasheets. Use of such speakers will increase the output power more than designed and the IC may pack up.

All said and done, IC are extremely rugged and forgiving over reasonable mistakes. I and most of the hobbyists have made mortal mistakes, while working on the projects and have removed ICs without any hope, feeling that they are gone. Disbelievingly they worked again. Many times we were not that lucky! But caution is better than crying on split waters.

Happy Experiments in Exciting Electronics! Ask for more!

SELF-IMPROVEMENT

New

9698 R • ₹ 195/- 9498 C • ₹ 180/- 9490 H • ₹ 175/- 9464 R • ₹ 80/- 9096 B • ₹ 150/- 5614 E • ₹ 150/- 4008 J • ₹ 150/- 9026 D • ₹ 120/- 9786 M • ₹ 195/-

9491 J • ₹ 100/- 8885 D • ₹ 150/- 9081 D • ₹ 150/- 9091 B • ₹ 120/- 9060 B • ₹ 195/- 9684 F • ₹ 195/- 8928 D • ₹ 80/- 9449 A • ₹ 195/- 9788 R • ₹ 195/-

MANAGEMENT/JOB/CARRIER/BUSINESS & PROFESSION

All Time Bestsellers

9461 K • ₹ 150/- 5338 A • ₹ 135/- (with CD) 8979 A • ₹ 135/- 9406 B • ₹ 150/- 9672 G • ₹ 150/- 9682 D • ₹ 120/- 8729 T • ₹ 120/-

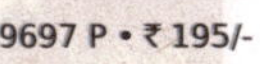

9697 P • ₹ 195/- 9313 D • ₹ 150/- 5623 B • ₹ 250/- 9439 L • ₹ 150/- 5441 D • ₹ 195/- 8883 D • ₹ 150/- 8735 F • ₹ 150/-

4018 D • ₹ 150/- 9079 B • ₹ 195/- 4005 E • ₹ 195/- 5643 B • ₹ 120/- 9431 C • ₹ 175/- 8990 C • ₹ 96/- 9763 P • Rs. 195/-

5618 D • ₹ 120/- 5640 C • ₹ 120/- 5615 D • ₹ 150/- 8972 C • ₹ 80/- 4001 A • ₹ 150/- 5646 A • ₹ 225/- 4017 D • ₹ 150/-

PERSONALITY DEVELOPMENT

8748 E • ₹ 195/-

9666 A • ₹ 150/-

9678 R • ₹ 195/-

9670 E • ₹ 240/-

9696 M • ₹ 220/-

9070 B • ₹ 195/-

9028 D • ₹ 175/-

5641 A • ₹ 150/-

9450 B • ₹ 195/-

9088 C • ₹ 195/-

9667 B • ₹ 150/-

8966 E • ₹ 100/-

5639 B • ₹ 80/-

9466 T • ₹ 96/-

9973 B • ₹ 110/-

9981 B • ₹ 96/-

8868 D • ₹ 120/-

9487 E • ₹ 150/-

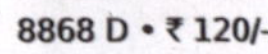

STUDENT DEVELOPMENT

9090 A • ₹ 220/-

9668 C • ₹ 150/-

9071 D • ₹ 165/-

8731 B • ₹ 100/-

9495 R • ₹ 175/-

9455 C • ₹ 150/-

5622 A • ₹ 120/-

9967 C • ₹ 120/-

2241 J • ₹ 100/-

94441 S • ₹ 195/-

9654 D • ₹ 100/-

9652 D • ₹ 120/-

8962 A • ₹ 150/-

9089 D • ₹ 135/-

4016 D • ₹ 160/-

4009 K • ₹ 150/-

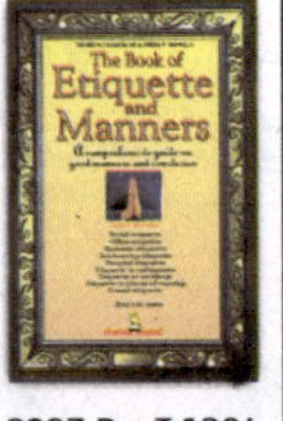

8997 B • ₹ 120/-

4010 L • ₹ 100/-

9787 P • ₹ 100/-

2244 D • ₹ 80/-

PARENTING

9906 J • ₹ 250/- (HB)

8261 D • ₹ 180/-

9674 J • ₹ 220/-

9784 J • ₹ 150/-

9594 K • ₹ 80/-

8917 D • ₹ 120/-

9458 G • ₹ 80/-

9438 B • ₹ 150/-

9065 A • ₹ 80/-

9994 E • ₹ 120/-

ALTERNATIVE THERAPIES

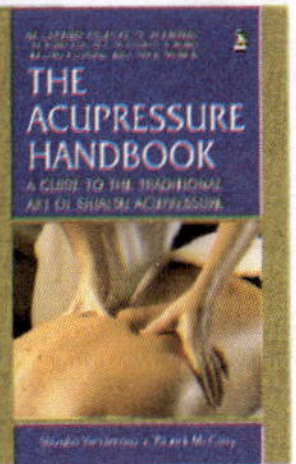

8882 F • ₹ 215/-

8983 E • ₹ 100/-

8836 D • ₹ 135/-

9935 F • ₹ 120/-

5637 D • ₹ 96/-

8889 D • ₹ 100/-

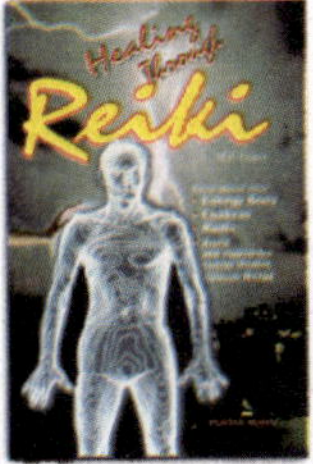

8842 D • ₹ 100/-

8941 A • ₹ 100/-

COMMON AILMENTS & DISEASES

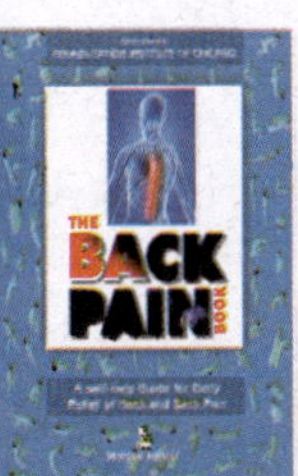

8891 D • ₹ 120/-

8281 A • ₹ 100/-

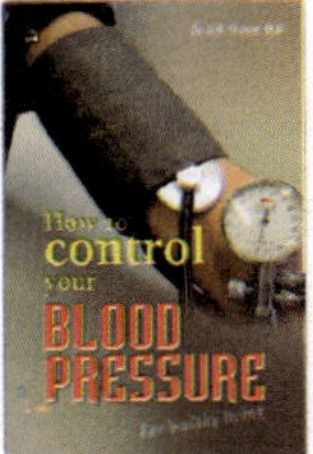

8094 D • ₹ 120/-

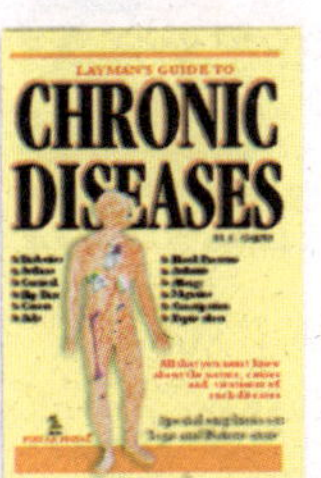

8848 D • ₹ 150/-

8276 A • ₹ 96/-

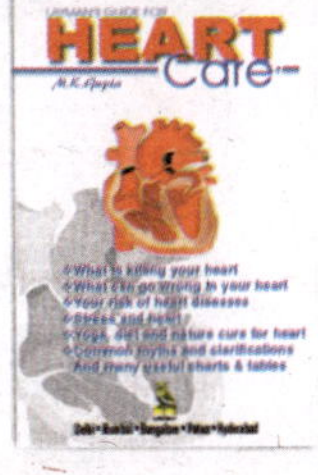

8888 D • ₹ 96/-

8908 D • ₹ 120/-

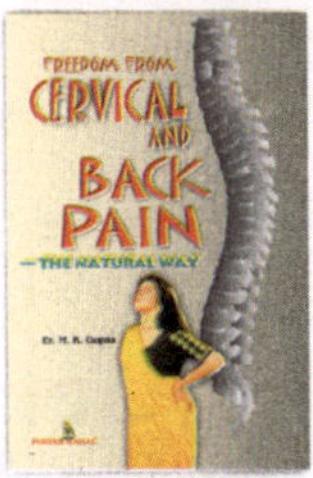

8878 B • ₹ 80/-

GENERAL HEALTH

9075 C • ₹ 225/-

8747 D • ₹ 150/-

9940 D • ₹ 150/-

8859 G • ₹ 80/-

8877 A • ₹ 150/-

8847 M • ₹ 100/-

8870 D • ₹ 100/-

9950 B • ₹ 120/-

9902 F • ₹ 120/-

SLIMMING & FITNESS

8277 B • ₹ 120/-

8875 K • ₹ 120/-

9445 A • ₹ 150/-

DIET & NUTRITION

9941 D • ₹ 100/-

8904 D • ₹ 150/-

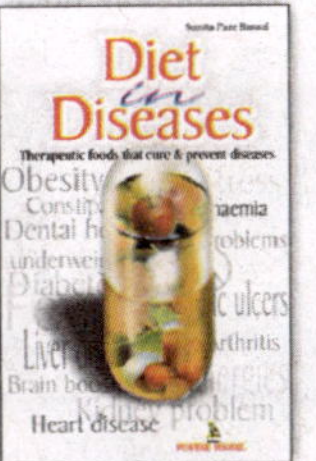

8985 B • ₹ 120/-

8968 G • ₹ 120/-

8271 C • ₹ 96/-

9037 D • ₹ 150/-

HINDOOLOGY / RELIGION / SPIRITUAL BOOKS

9873 C • ₹ 60/-

9770 E • ₹ 150/-

9453 A • ₹ 250/-

4179 A • ₹ 295/- (HB)

4138 B Rs. 250

4177 B • ₹ 250/-

9997 C • ₹ 80/-

4181 C • ₹ 195/-

9984 E • ₹ 399/- (HB)

4130 B • ₹ 120/-

9811 P • ₹ 120/-

9585 A • ₹ 96/-

9508 D • ₹ 95/-

9989 D • ₹ 96/-

4183 A • ₹ 350/- (HB)

9504 D • ₹ 100/-

9540 D • ₹ 150/-

9513 A • ₹ 195/-

4126 B • ₹ 96/-

9812 R • ₹ 120/-

9504 D • ₹ 100/-

4124 A • ₹ 120/-

4190 C • ₹ 160/-

9509 A • ₹ 150/-

4152 B • ₹ 96/-

4188 A • ₹ 160/-

4132 D • ₹ 100/-

9987 E • ₹ 150/-

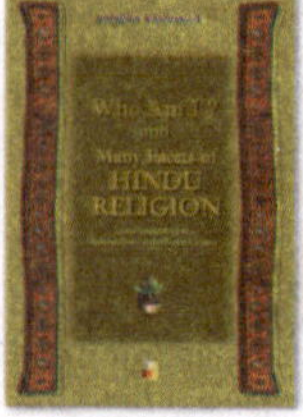

9520 D • ₹ 120/-

4134 B • ₹ 80/-

4182 D • ₹ 96/-

9405 A • ₹ 19

COMPUTERS

7712 K • ₹ 165/-

7711 J • ₹ 12(

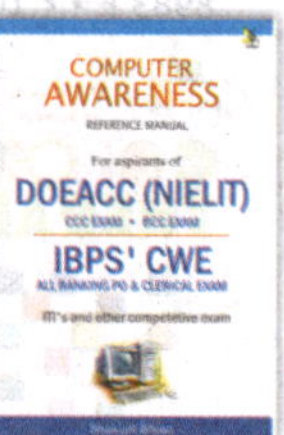

9768 C • ₹ 175/-

7766 A • ₹ 12(

HOME MAKING / GRILLS & RAILIN

3111 E • ₹ 175/-

3107 F • ₹ 88

3106 E • ₹ 100/-

3105 D • ₹ 10(

3108 G • ₹ 150/-

3104 M • ₹ 10(